Statistics

A course for A level Mathematics

Book 2

M. E. M. Jones

Head of the Mathematics Department at the Howardian High School, Cardiff

Schofield & Sims Ltd Huddersfield

0 7217 2361 6
0 7217 2365 9 Net Edition

First printed 1989

Acknowledgements

The author and publishers are grateful for permission to use
questions from past G.C.E. examinations. These are
acknowledged as follows:

> Joint Matriculation Board (JMB)
> Welsh Joint Education Committee (WJEC)

Grateful acknowledgement is also due to the Joint
Matriculation Board for permission to reproduce the
statistical tables on pages 229 to 233 from their booklet
S159(A).

The Examining Boards whose questions are reproduced bear
no responsibility whatever for the answers to examination
questions given here, which are the sole responsibility of
the author.

Designed by Graphic Art Concepts, Leeds
Printed in Great Britain at the Alden Press, Oxford

Author's note

This book, together with Book 1, covers the statistics content of the syllabuses in Advanced Level Mathematics (Pure Mathematics with Statistics) and Further Mathematics (Pure and Applied Mathematics) of the Joint Matriculation Board, and in Advanced Level Mathematics and Applied Mathematics of the Welsh Joint Education Committee. The book aims to give a sound introduction to probabilistic and statistical concepts which are relevant to other school courses at both AS and A level, and also to introductory courses at colleges and universities.

At the time of publication, the book covers more topics than appear in any one of the aforementioned syllabuses. Readers who are following either of the JMB syllabuses may omit sections **3.1**, **3.2**, **7.6** and **8.4**; in addition, Further Mathematics candidates may omit the whole of Chapter 9. Readers following the WJEC syllabuses may omit the whole of Chapter 8.

Each chapter contains numerous worked examples to illustrate the methods by which the relevant theory is applied, followed by routine exercises to consolidate progress. At the end of each chapter is a miscellaneous exercise containing some more demanding questions most of which are taken from past papers of the Joint Matriculation Board (JMB) and Welsh Joint Education Committee (WJEC).

I wish to express my grateful thanks to a number of people who have assisted in the production of this book. In particular, to Dr. I.G. Evans, for reading the whole book in the early stages of its production and for making many helpful comments. Thanks also to the staff of Schofield and Sims for their invaluable help. Finally, thanks to my wife, Joan, for her help and encouragement and to my son, Christopher, for help with computing and word-processing.

M. E. M. Jones

Contents

Chapter 1

Poisson distribution

Chapter 2

The distribution of a function of a random variable

Chapter 3

Joint distributions of discrete random variables

Chapter 4

Joint distributions of continuous random variables

Chapter 5

Sampling distributions

Chapter 6

Estimation

iv

Chapter 7

Confidence intervals

Chapter 8

Hypothesis testing

Chapter 9

Linear relationships

Chapter 10

Revision

Notation

S	the sample space, the set of all possible outcomes of a random experiment
$A, B, C, \ldots$	subsets of S, events $A, B, C, \ldots$
$P(A)$	the probability of the event A
$P(A \mid B)$	the conditional probability of the event A given that the event B has occurred
$X, Y, Z, \ldots$	random variables $X, Y, Z, \ldots$
x_i	an arbitrary value of a discrete random variable X
p_i	the probability $P(X = x_i)$
$p(x_i)$	the value of the probability function p of a discrete random variable X
x	an arbitrary value of a continuous random variable X
$f(x)$	the value of the probability density function f of a continuous random variable X
$F(x)$	the value of the cumulative distribution function F of a continuous random variable X
$E[X]$	the expected value or expectation of a random variable X
$E[g(X)]$	the expected value of $g(X)$
$V[X]$	the variance of a random variable X
$SD[X]$	the standard deviation of a random variable X
μ	a population mean
σ^2	a population variance
σ	a population standard deviation
$\bar{x}$	a sample mean
s^2	a sample unbiased estimate of a population variance
$\sim$	is distributed as
$B(n, p)$	the binomial distribution with index n and probability parameter p
$Po(\alpha)$	the Poisson distribution with mean α
$U(a, b)$	the continuous uniform distribution over the interval (a, b)
$N(\mu, \sigma^2)$	the normal distribution with mean μ and variance σ^2
Z	the random variable having the distribution $N(0, 1)$
ϕ	the probability density function of Z
Φ	the cumulative distribution function of Z

Chapter 1

Poisson distribution

In Chapter 6 of Book 1 probability distributions of discrete random variables were introduced, and in Chapter 7 some special distributions, such as the binomial distribution and the geometric distribution were studied. In this chapter another very important discrete distribution, the Poisson distribution, will be introduced.

1.1 Properties of a Poisson distribution

A discrete random variable X which has a distribution given by

$$P(X = r) = e^{-\alpha}\frac{\alpha^r}{r!}, \qquad r = 0, 1, 2, \ldots$$

where $\alpha > 0$, is said to have a Poisson distribution with parameter α. This is often abbreviated as $X \sim Po(\alpha)$.

It may be shown (using mathematics beyond A level) that the Poisson distribution, with some value of α, is a reasonable model for the distribution of the number of occurrences in a fixed interval of time (or space) of an event which occurs at random in time (or space), provided that:

1 simultaneous occurrences of the event are not possible;

2 the mean number of occurrences of the event in intervals of a specific size is directly proportional to the size of the interval;

3 the occurrences of the event are independent.

Some examples of random variables which, in some circumstances, might have a Poisson distribution are given below.

a The number of telephone calls taken by a switchboard in a minute.

b The number of claims made to an insurance company in a month.

c The number of particles emitted by a radioactive source in half a minute.

d The number of flaws in a metre length of material.

e The number of gnat larvae in $10\,cm^3$ of pond water.

f The number of printing errors on a page of a proof copy of a book.

Probability generating function of a Poisson distribution

Probability generating functions of discrete random variables were introduced on page 114 of Book 1 and the results, quoted below, concerning the mean and the variance were established on pages 114 and 115.

Consider the function G given by

$$G(t) = e^{\alpha(t-1)}$$

where α is a positive constant.

$$G(t) = e^{-\alpha} \cdot e^{\alpha t}$$

$$= e^{-\alpha}\left(1 + \alpha t + \frac{\alpha^2 t^2}{2!} + \cdots + \frac{\alpha^r t^r}{r!} + \cdots\right)$$

The coefficient of t^r in the expansion of $G(t)$ is $e^{-\alpha}\dfrac{\alpha^r}{r!}$, which is $P(X = r)$ and therefore G is the probability generating function for Po(α).

Mean and variance of a Poisson distribution

$$G(t) = e^{\alpha(t-1)}$$

Differentiating with respect to t $\quad G'(t) = \alpha e^{\alpha(t-1)}.$

Differentiating again with respect to t $\quad G''(t) = \alpha^2 e^{\alpha(t-1)}.$

But the mean μ is given by $\quad \mu = G'(1)$ (page 114, Book 1)

$$= \alpha e^0.$$

Therefore $\quad \mu = \alpha.$

Also the variance σ^2 is given by $\quad \sigma^2 = G''(1) + \mu - \mu^2$ (page 115, Book 1)

$$= \alpha^2 e^0 + \alpha - \alpha^2.$$

Therefore $\quad \sigma^2 = \alpha.$

The distribution Po(α) has mean α and variance α.

Example 1

The number of accidents in a week in a factory has a Poisson distribution with mean 4. Find the probability that the number of accidents that will occur in a week is less than four.

Let X be the number of accidents in a week.

$$X \sim \text{Po}(4)$$

$$P(X < 4) = P(X = 0) + P(X = 1) + P(X = 2) + P(X = 3)$$

$$= e^{-4} \cdot \frac{4^0}{0!} + e^{-4} \cdot \frac{4^1}{1!} + e^{-4} \cdot \frac{4^2}{2!} + e^{-4} \cdot \frac{4^3}{3!}$$

$$= e^{-4}\left(1 + 4 + 8 + \frac{32}{3}\right)$$

$$= 0.433\,(3\text{ d.p.})$$

Recursive formula

The calculation of Poisson probabilities may be facilitated by means of the formula

$$\frac{p_r}{p_{r-1}} = \frac{\alpha}{r}$$

where $p_r = P(X = r)$ and $p_{r-1} = P(X = r - 1)$.

Proof

$$\frac{p_r}{p_{r-1}} = e^{-\alpha}\frac{\alpha^r}{r!} \div e^{-\alpha}\frac{\alpha^{r-1}}{(r-1)!}$$

$$= e^{-\alpha}\frac{\alpha^r}{r!} \times \frac{(r-1)!}{e^{-\alpha}\alpha^{r-1}}$$

$$= \frac{\alpha}{r}$$

Using the recursive formula the calculation of $P(X < 4)$ in **Example 1** above may be performed on a calculator as follows.

$$p_0 = e^{-4} = 0.0183156$$

Put this into the memory. Do not clear the display. Use the recursive formula with $r = 1$ and $\alpha = 4$.

$$p_1 = \frac{4}{1} \times p_0 = 0.0732625$$

Add this to the memory. Do not clear the display. Use the recursive formula with $r = 2$ and $\alpha = 4$.

$$p_2 = \frac{4}{2} \times p_1 = 0.1465251$$

Add this to the memory. Do not clear the display. Use the recursive formula with $r = 3$ and $\alpha = 4$.

$$p_3 = \frac{4}{3} \times p_2 = 0.1953668$$

Add this to the memory. Recall the memory to give the required answer.

$$P(X < 4) = 0.433\,(3\,\text{d.p.})$$

Tables of cumulative Poisson probabilities

Table 2 (page 230) tabulates values of $P(X \leqslant r)$, where $X \sim Po(\alpha)$, for certain values of α and r. In suitable examples the use of **Table 2** can reduce the amount of calculation required.

In **Example 1** above, $P(X < 4)$ is equal to $P(X \leqslant 3)$ which may be read directly from **Table 2** (in the column labelled $a = 4.0$ and the row labelled $r = 3$).

Other probabilities associated with Po(α) may be found as follows.

$$P(X = r) = P(X \leqslant r) - P(X \leqslant r - 1)$$
$$P(X < r) = P(X \leqslant r - 1)$$
$$P(X \geqslant r) = 1 - P(X \leqslant r - 1)$$
$$P(X > r) = 1 - P(X \leqslant r)$$

Example 2

The number of cars arriving per hour at a petrol station has a Poisson distribution with mean 27. Find the probability that
a 30 cars will arrive during an hour
b more than 2 cars will arrive during a 5-minute interval
c less than 7 cars will arrive during a 20-minute interval.

a Let X be the number of cars arriving in an hour.

$$X \sim Po(27)$$
$$P(X = 30) = e^{-27}\frac{27^{30}}{30!}$$
$$= 0.062 \text{ (3 d.p.)}$$

b Let Y be the number of cars arriving in a 5-minute interval. From the properties of the Poisson distribution listed above, it follows that Y has a Poisson distribution with a mean of $\frac{27}{12}$.

$$Y \sim Po(2.25)$$

Since $a = 2.25$ is not listed in **Table 2**, page 230, the probabilities required have to be evaluated on a calculator.

$$P(Y > 2) = 1 - P(Y \leqslant 2)$$
$$= 1 - (P(Y = 0) + P(Y = 1) + P(Y = 2))$$
$$= 1 - \left(e^{-2.25} + e^{-2.25}.2.25 + e^{-2.25}.\frac{2.25^2}{2!}\right)$$
$$= 0.391 \text{ (3 d.p.)}$$

c Let Z be the number of cars arriving in a 20-minute interval. Z has a Poisson distribution with mean $\frac{27}{3}$.

$$Z \sim Po(9)$$
$$P(Z < 7) = P(Z \leqslant 6)$$
$$= 0.207 \quad \text{(3 d.p.)} \qquad \text{(from \textbf{Table 2})}$$

Exercise 1.1

1 Given that $X \sim Po(3)$ determine the values of
  a $P(X = 5)$ **b** $P(X < 4)$ **c** $P(X > 2)$.

2 Given that $X \sim Po(8.5)$ determine the values of
  a $P(X = 7)$ **b** $P(X \leqslant 11)$ **c** $P(X \geqslant 4)$.

3 Given that $X \sim$ Po(6) determine the values of
    a $P(X = 4)$ **b** $P(X < 7)$ **c** $P(X \geqslant 10)$.

4 Given that $X \sim$ Po(7.4) calculate the values of
    a $P(X = 6)$ **b** $P(X \leqslant 4)$ **c** $P(X > 3)$.

5 Given that $X \sim$ Po(12) calculate the values of
    a $P(X = 8)$ **b** $P(7 < X < 13)$ **c** $P(X > 4)$.

6 The number of flaws in a metre length of material produced by a certain machine has a Poisson distribution with mean 0.5. Find the probability that
    a   a metre length has no flaw,
    b   a metre length has more than one flaw,
    c   a 10-metre length has no flaw,
    d   a 10-metre length has fewer than four flaws.

7 The number of incoming calls per hour to a switchboard has a Poisson distribution with mean 10. Find the probability that the number of incoming calls to the switchboard in a particular hour
    a   will be 10,
    b   will exceed 9,
    c   will be less than 5.

8 The number of insurance claims made per day on an insurance company has a Poisson distribution with mean 5. Find the probability that the number of claims made on a particular day
    a   will be 3,
    b   will exceed 3,
    c   will be less than 6.

9 The number of dragonfly larvae in 1 cl of water from a certain pond has a Poisson distribution with mean 0.3. Four independent samples, each 1 cl, of the pond water are taken. Find the probability that
    a   none of the samples contain larvae,
    b   one sample contains one larva and the others contain no larvae.
    c   Find the expected value of the number of samples containing no larvae.

10 The number of accidents occurring in a week in a certain factory has a Poisson distribution with a standard deviation of 1.9. Find the probability that the number of accidents in the factory in a particular week
    a   will be 8,
    b   will be less than 4,
    c   will be more than 2.

1.2 Poisson approximation to a binomial distribution

Given that $X \sim B(n, p)$, where n is large and p is small, it may be shown that the distribution of X may be approximated by the distribution of a random variable Y having a Poisson distribution with mean $\alpha = np$. There is no simple rule which indicates how large n must be and how small p must be for the approximation to be good, but some guidance is given below.

Cumulative probabilities associated with $B(100, 0.01)$, $B(50, 0.08)$ and their Poisson approximations $Po(1)$, $Po(4)$ are listed in the following table.

r	$X \sim B(100, 0.01)$ $P(X \leqslant r)$	$Y \sim Po(1)$ $P(Y \leqslant r)$	$X \sim B(50, 0.08)$ $P(X \leqslant r)$	$Y \sim Po(4)$ $P(Y \leqslant r)$
0	0.366	0.368	0.015	0.018
1	0.736	0.736	0.083	0.092
2	0.921	0.920	0.226	0.238
3	0.982	0.981	0.425	0.433
4	0.997	0.996	0.629	0.629
5	1.000	0.999	0.792	0.785

It is clear that the Poisson approximation is very good for $n = 100$, $p = 0.01$ and that it is moderately good for $n = 50$, $p = 0.08$.

Empirical studies* indicate that, for $p < 0.5$, the maximum absolute error in the cumulative probabilities is nearly independent of n and is approximately $0.2p$. Thus when $p < 0.1$ the Poisson approximation to the binomial distribution is usually satisfactory (unless the probability being approximated is itself small, when the percentage error may be large).

Example 3

The mortality rate from a certain disease is 0.7%. Find the probability that there will be fewer than 6 deaths from the disease in a group of 500 people.

Let X be the number of deaths in a group of 500.

X is binomially distributed with $n = 500$ and $p = 0.007$.

Let Y have Poisson distribution with mean $500 \times 0.007 = 3.5$.

$$P(X < 6) \simeq P(Y \leqslant 5)$$
$$= 0.858 \quad (3 \text{ d.p.}) \qquad (\text{from } \textbf{Table 2}, \text{ page 230})$$

This is exactly the same as the value (to 3 d.p.) calculated directly from the binomial distribution.

The Poisson approximation may also be used for evaluating a binomial probability when p is close to 1 and n is large. In this case the distribution of the number of failures is approximated by a Poisson distribution.

* (see 'The error in approximating cumulative binomial and Poisson probabilities' by J. Green and J. Round-Turner in *Teaching Statistics*, Vol. 8, No. 2, May 1986)

Example 4

The probability that a particular type of flower seed will germinate under certain planting conditions is 0.98. If 100 seeds are planted under these conditions use an approximate method to calculate the probability that at least 95 will germinate.

Let X be the number of seeds germinating, then $X \sim B(100, 0.98)$.

Let X' be the number of seeds not germinating, then $X' \sim B(100, 0.02)$.

Let Y have a Poisson distribution with mean $100 \times 0.02 = 2$. $Y \sim Po(2)$.

$$P(X \geqslant 95) = P(X' \leqslant 5)$$
$$\simeq P(Y \leqslant 5)$$
$$= 0.983 \quad (3\,d.p.) \quad \text{(from \textbf{Table 2}, page 230)}$$

This compares with the answer 0.985 (to 3 d.p.) obtained from direct calculation of the binomial probabilities.

Exercise 1.2

1 Given that $X \sim B(100, 0.04)$ find to 3 decimal places approximate values for

   a $P(X \leqslant 2)$,   **b** $P(X < 5)$.

Compare these values with those obtained by using the binomial probabilities.

2 A unit trust receives buying orders, on average, from 98% of persons who make enquiries in response to newspaper advertisements. If 200 persons respond to the latest advertisement find an approximate value for the probability that at least 195 of them will place buying orders.

3 The probability that a typist will mistype a character is 0.005. When the typist types a page containing 480 characters find, to 3 decimal places, the probability that the number of typing mistakes will exceed 5.

4 The probability that an electronic component is faulty is 0.003. These components are packed in boxes of 500. Using a suitable approximation, calculate the probability that two boxes selected at random will contain in total at most 5 faulty components.

5 An aeroplane has 296 seats. From past experience the airline knows that, on average, 1.5% of passengers with tickets for a particular flight fail to arrive for that flight. If the airline sells 300 tickets for a flight, find approximate values for

   a the probability that the flight will be overbooked,

   b the probability that there will empty seats on the flight.

1.3 Normal approximation to a Poisson distribution

When $X \sim \text{Po}(\alpha)$, where α is large, it may be shown that the distribution of X may be approximated by the normally distributed random variable Y with mean α and variance α.

The formal justification of this statement is too difficult at this stage but the diagrams below indicate graphically that the approximation is moderately good when $\alpha = 10$ and is good when $\alpha = 20$. Each diagram shows a theoretical relative frequency histogram for $\text{Po}(\alpha)$ with the graph of the probability density function of $N(\alpha, \alpha)$ superimposed on it.

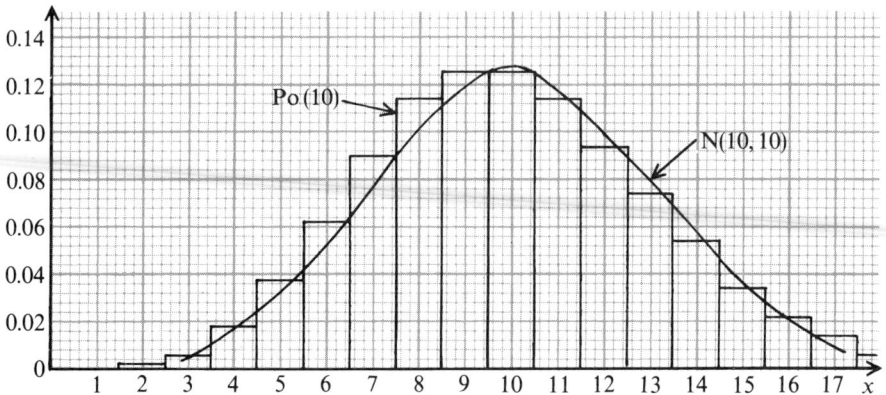

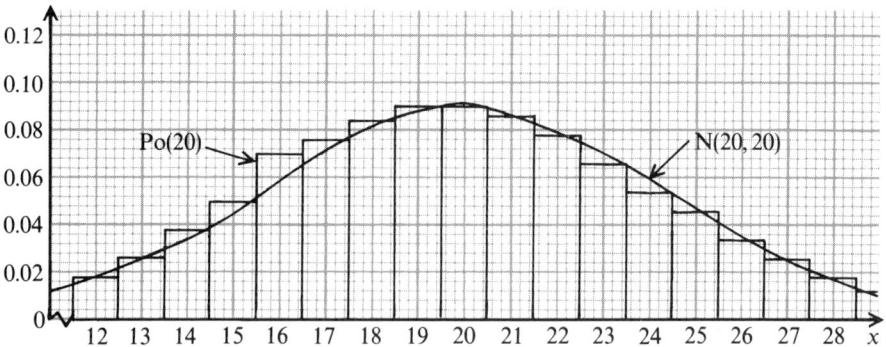

Continuity corrections

Since a Poisson distribution is discrete and its normal approximation is continuous, it is necessary to apply the following continuity corrections when a normal approximation is used to facilitate the calculation of Poisson probabilities.

$$P(X = r) \simeq P(r - 0.5 < Y < r + 0.5)$$

$$P(X \leqslant r) \simeq P(Y < r + 0.5)$$

$$P(X \geqslant r) \simeq P(Y > r - 0.5)$$

As a rule of thumb the normal approximation to Po(α) may be used whenever $\alpha > 16$, justified below.

The range of values of X is from 0 to $+\infty$ whilst the range of values of Y is from $-\infty$ to $+\infty$. For the approximation to be good it is necessary that $P(0 < Y < +\infty)$ should close to 1. Since the probability that a normal random variable, having mean μ and standard deviation σ, lies between $\mu - 4\sigma$ and $\mu + 4\sigma$ is 0.99994 it is desirable that

$$\alpha - 4\sqrt{\alpha} > 0$$
$$\sqrt{\alpha} > 4$$
$$\alpha > 16.$$

When using the normal approximation to a Poisson distribution the maximum error in the cumulative probabilities is less than 0.017 for $\alpha = 16$; the maximum error decreases as α increases and is less than 0.01 for values of α greater than 50.

Example 5

The number of accidents in a factory has a Poisson distribution. The mean number of accidents per month is 3. Calculate, to three decimal places, an approximate value for the probability that the number of accidents in a year

a   will exceed 25

b   will lie between 26 and 46 inclusive.

Let X be the number of accidents in a year. The distribution of X is Poisson with mean $3 \times 12 = 36$.

Let $Y \sim N(36, 36)$, then $Z = \dfrac{(Y - 36)}{6} \sim N(0, 1)$.

a $P(X > 25) \simeq P(Y > 25.5)$

$$\simeq P\left(Z > \frac{25.5 - 36}{6}\right)$$

$$\simeq P(Z > -1.75)$$

$$\simeq 0.960 \quad \text{(3 d.p.)} \qquad \text{(from \textbf{Table 3}, page 231)}$$

This compares with the value 0.965 calculated using Poisson probabilities.

b $P(26 \leqslant X \leqslant 46) \simeq P(25.5 < Y < 46.5)$

$$\simeq P\left(\frac{25.5 - 36}{6} < Z < \frac{46.5 - 36}{6}\right)$$

$$\simeq P(-1.75 < Z < 1.75)$$

$$\simeq 0.920 \quad \text{(3 d.p.)} \qquad \text{(from \textbf{Table 3})}$$

This compares with the value 0.921 calculated using Poisson probabilities.

STATISTICS 2

Exercise 1.3

1 Given that $X \sim \text{Po}(40)$, use a normal approximation to find
   a $\text{P}(X \leqslant 30)$ b $\text{P}(35 < X < 45)$ c $\text{P}(X \geqslant 40)$.

2 Given that $X \sim \text{Po}(100)$, use a normal approximation to find
   a $\text{P}(X > 96)$ b $\text{P}(90 \leqslant X \leqslant 102)$ c $\text{P}(X < 110)$.

3 The number of telephone calls received by an office switchboard during a certain hour of the working day has a Poisson distribution with mean 24. Use a normal approximation to calculate the probability that, in that hour of a particular working day,
   a there will be fewer than 20 calls,
   b there will be more than 30 calls,
   c there will be exactly 24 calls.

4 The number of particles emitted per second from a radioactive source has a Poisson distribution with mean 50. Calculate the probability that in one second the number of particles emitted lies between 48 and 52 inclusive,
   a using the Poisson distribution,
   b using the normal approximation to the Poisson distribution.

5 The number of eggs laid by blackflies has a Poisson distribution with mean 150. Calculate the probability that a blackfly
   a will lay more than 160 eggs,
   b will lay fewer than 150 eggs.

1.4 Further examples

Example 6

In checking twenty pages of a draft copy of a document it was found that only one page contained no error. Assuming that the number of errors per page has a Poisson distribution with mean α, estimate the value of α to the nearest whole number.
If the document contains 500 pages, estimate the number of pages containing more than six errors.

Let X be the number of errors per page.
$$X \sim \text{Po}(\alpha)$$
Given that $\qquad \text{P}(X = 0) = \dfrac{1}{20}$
Therefore $\qquad e^{-\alpha} = 0.05$
$$-\alpha = \ln(0.05)$$
Estimated value of $\qquad \alpha = 3$ (nearest whole number)

10

$$P(X \leqslant 6) = 0.966 \qquad \text{(from \textbf{Table 2}, page 230)}$$
$$P(X > 6) = 1 - 0.966$$
$$= 0.034$$

Estimated number of pages $= 0.034 \times 500$
$$= 17$$

Example 7

A small car-hire firm has three cars available for daily hire. The daily demand for these cars has a Poisson distribution with mean 2.5.
a Calculate the mean number of cars hired per day.
b The firm charges £20 per day for the hire of each car and has expenses of £5 per car per day whether the cars are hired or not. Calculate the firm's expected daily profit from the hire of these cars.

a Let X be the daily demand, then $X \sim \text{Po}(2.5)$.
Let Y be the number of cars hired per day.

$$P(Y = 0) = P(X = 0) = 0.082 \qquad \textbf{(Table 2)}$$
$$P(Y = 1) = P(X = 1) = 0.287 - 0.082 = 0.205 \qquad \textbf{(Table 2)}$$
$$P(Y = 2) = P(X = 2) = 0.544 - 0.287 = 0.257 \qquad \textbf{(Table 2)}$$

However, when the daily demand is equal to 3 or greater than 3, the number of cars hired is 3, since this is the largest number of cars available for hire.
$$P(Y = 3) = P(X \geqslant 3) = 1 - 0.544 = 0.456 \qquad \text{(from \textbf{Table 2})}$$

The probability distribution of Y is given by the first two columns of the following table and the mean is calculated by evaluating $\sum p_i y_i$.

y_i	p_i	$p_i y_i$
0	0.082	0
1	0.205	0.205
2	0.257	0.514
3	0.456	1.368
	1.000	2.087

Check that $\sum p_i = 1$ (apart from a possible rounding error)
$$E[Y] = 2.087.$$

The mean number of cars hired per day is 2.09   (2 d.p.)

b Let £P be the daily profit from the hire of these cars.
$$E[P] = E[20Y - 15]$$
$$= 20E[Y] - 15$$
$$= 20 \times 2.087 - 15$$
$$= 26.74$$

The expected daily profit is £26·74.

Example 8

The number of loaded oil tankers arriving at a port between successive high tides has a Poisson distribution with mean 2. Loaded tankers can enter the dock area only at high tide. The port has dock space for only three tankers, which are discharged and leave the dock area before the next high tide. Only the first three loaded tankers waiting at any high tide go into the dock area; any others must await the next high tide.

Starting from an evening high tide after which no ship remains waiting its turn, find the probability that no tanker is left waiting outside the dock area

a after the following morning's high tide,

b after the following evening's high tide.

a Let X be the number of tankers arriving before the next morning's high tide.

$$X \sim Po(2)$$

$$P(\text{required}) = P(X \leqslant 3)$$

$$= 0.857 \quad (\text{from **Table 2**, page 230})$$

b Let Y be the number of tankers arriving between the next morning's high tide and the following evening's high tide.

$$Y \sim Po(2)$$

Assuming that X and Y are independent, the probability p required is given by:

$$p = P(X \leqslant 3).P(Y \leqslant 3) + P(X = 4).P(Y \leqslant 2) + P(X = 5).P(Y \leqslant 1) + P(X = 6).P(Y = 0).$$

Using **Table 2** to evaluate the probabilities on the right hand side.

$$p = 0.857 \times 0.857 + 0.090 \times 0.677 + 0.036 \times 0.406 + 0.012 \times 0.135$$
$$= 0.81 \quad (2\,\text{d.p.})$$

Miscellaneous Exercise ☐ 1 ☐

1 In each trial of a certain experiment the probability of an event A occurring is p. Let X denote the number of times that A will occur in 100 independent trials of the experiment. In each of the following cases, use a suitable approximate method to evaluate $P(X \geqslant m)$ correct to three decimal places:

(i) $p = 0.6$ and $m = 69$,

(ii) $p = 0.01$ and $m = 4$. *(JMB)*

2 Define the Poisson distribution and derive its mean and variance. The number of particles emitted from a radioactive source in t seconds has Poisson distribution with mean $\dfrac{t}{20}$. Find, correct to two decimal places, the probabilities that in a period of one minute

(i) no particle,

(ii) at least three particles,

will be emitted.

Use the normal approximation to find the probability, correct to two decimal places, that at least 200 particles will be emitted in one hour.

(*JMB*)

3 State the conditions under which it is permissible to use the Poisson distribution as an approximation to the binomial distribution.

It is known that 0.6% of the components produced by a factory are defective. Each day a random sample of 200 components is inspected. Find the probability that there are no defectives in this daily sample

(i) using the binomial distribution,

(ii) using the Poisson distribution.

Find, to two decimal places, the probability that there is at least one defective on each of three successive days. Taking the components inspected on the three successive days as a single sample, use the Poisson distribution to calculate, to two decimal places, the probability of three or more defectives in the three days. State briefly why you would expect the second of the last two probabilities to be greater than the first. (*JMB*)

4 A factory has three machines A, B and C producing a certain type of item. Of the daily output, 50% of the items are produced on A, 30% on B and 20% on C. The probabilities that an item produced on A, B and C is defective are 0.02, 0.02 and 0.07, respectively.

a One item is chosen at random from a day's total output.

(i) Show that the probability that it will be defective is 0.03.

(ii) Given that the chosen item is defective, calculate the probability that it was produced on machine C.

b A random sample of n items is chosen from a day's total output, which may be assumed to be very large.

(i) If $n = 100$, find the probability that the sample will include exactly 3 or 4 defective items.

(ii) If $n = 500$, find an approximate value for the probability that the sample will include 5 or fewer defective items. (*WJEC*)

5 a A contestant in a television quiz programme is required to answer eight true/false questions (that is, the contestant has to indicate whether the statement made in the question is true or false). If a contestant guesses the answer to each of the eight questions, calculate the probability that he will obtain at least six correct answers.

b In each trial of a random experiment the probability that the event A will occur is 0.02.

(i) Find, correct to three decimal places, the probability that in twenty independent trials of the experiment the event A will occur not more than once.

(ii) If 500 independent trials of the experiment are performed, use an appropriate approximate procedure to find, correct to three decimal places, the probability that the event A will occur exactly five times. *(WJEC)*

6 Each play of a one-armed bandit machine has a probability 0.04 of yielding a win. It may be assumed that this probability applies independently for each play of the machine.

a Find the probability that in ten plays of the machine there will be
(i) at least two wins,
(ii) exactly two wins.

b For a sequence of n plays of the machine write down an expression for the probability that there will be no win. Hence find the smallest number of plays of the machine for there to be a probability of at least 0.95 of winning at least once.

c Use an appropriate method to find an approximate value for the probability that there will be six or fewer wins in 80 plays of the machine. *(WJEC)*

7 A garage owner has three cars available for self-drive hire. The daily demand for these cars has a Poisson distribution with mean 2.

a Calculate, to two significant figures, the probabilities that on any day
(i) at least one car will not be hired,
(ii) the demand will exceed the number of cars available.

b Calculate, to three significant figures, the expected daily number of unhired cars.

c The garage owner charges £4 per day for the hire of a car, and his total outgoings per car, irrespective of whether or not it is hired, amount to £1 per day. Calculate, to the nearest penny, the garage owner's expected daily profit from the hiring of these cars. *(JMB)*

8 The monthly demand for a certain magazine at a small newsagent's shop has a Poisson distribution with mean 3. The newsagent always orders 4 copies of the magazine for sale each month; any demand for the magazine in excess of 4 is not met.

(i) Calculate the probability that the newsagent will not be able to meet the demand in a given month.

(ii) Find the most probable number of magazines *sold* in one month.

(iii) Find the expected number of magazines sold in one month.

(iv) Determine the least number of copies of the magazine that the newsagent should order each month so as to meet the demand with a probability of at least 0.95. *(JMB)*

9 A small garage has three cars available for daily hire. The daily demand for these cars may be assumed to have a Poisson distribution with mean 2.

(i) Prove that the demands for one and two cars on any day are equally probable.

(ii) Find, as accurately as your tables permit, the probabilities that on a given day exactly 0, 1, 2 and 3 cars will be hired.

Hence, or otherwise, calculate the mean number of cars hired per day.

(iii) The garage-owner charges £6 per day for the hire of a car and his total outgoings per car, irrespective of whether or not the car is hired, amount to £1 per day. Calculate, to the nearest penny, the garage-owner's expected daily profit from the hire of these cars.

(WJEC)

10 A newsagent has a regular order for 15 copies of a certain weekly magazine. Each week she sells 10 copies to regular customers and the weekly demand for the remaining copies may be assumed to have a Poisson distribution with mean 4. Let X denote the number of copies sold in a week. Find the probabilities with which X takes the values 10, 11, 12, 13, 14 and 15. Hence calculate the expected value of X.

The newsagent makes a profit of 30p on each copy sold and a loss of 10p on each copy unsold.

(i) Find the expected value of her net weekly profit from the sale of this magazine.

(ii) Determine whether it would be advisable for her to reduce her regular order to 14 copies, assuming that this would not change the distribution of the weekly demand. *(JMB)*

11 The number of flaws per roll of manufactured material has the Poisson distribution with mean 0.5.

(i) Find the probability that two randomly chosen rolls will have no flaw.

(ii) Five rolls are chosen at random one after the other. Find the probability that the fifth roll chosen is the only one which contains one or more flaws.

(iii) Find the largest number of rolls that can be chosen at random for there to be a probability of at least 0.2 that none of the rolls contains a flaw. *(JMB)*

12 A car ferry leaves a jetty at 10-minute intervals and can carry 2 cars. Cars arrive to board the ferry randomly and independently at an average rate of 12 per hour. Assuming that the number of arrivals in any 10-minute period has a Poisson distribution, and given that no cars are left behind at the first departure of the ferry, find the probabilities that

(i) no cars are left behind at the second departure,

(ii) no cars are left behind at the third departure,

(iii) just one car is left behind at the third departure. *(JMB)*

15

13 Define the Poisson distribution and derive its mean. State the circumstances under which it is appropriate to use the Poisson distribution as an approximation to the binomial distribution.

A lottery has a very large number of tickets, one in every 500 of which entitles the purchaser to a prize. An agent sells 1000 tickets for the lottery. Using the Poisson distribution, with the aid of the statistical tables provided or otherwise, find, to three decimal places, the probabilities that the number of prize-winning tickets sold by the agent is

(i)   less than three,

(ii) more than five.

Calculate the minimum number of tickets the agent must sell to have 95% chance of selling at least one prize-winning ticket. (*JMB*)

14 Define the Poisson distribution and derive its mean and variance.

In the first year of the life of a certain type of machine, the number of times a maintenance engineer is required has a Poisson distribution with mean four. Find the probability that more than four calls are necessary. The first call is free of charge and subsequent calls cost £20 each. Find the mean cost of maintenance in the first year. (*JMB*)

15 During a period of t months the demand for a particular item stocked in a shop is a Poisson random variable having mean $12t$.

(i)   Find the probability that the demand in one month will be in the range 10 to 20 inclusive.

(ii) The shop receives delivery of the items at the beginning of each month, when it accepts as many items as are necessary to bring the stock level up to n items. Find the smallest value of n for there to be a probability of at most 0.05 that the shop will not be able to meet the demand in a month.

(iii) Suppose that the shop acquires extra storage, so it can stock a very large number of these items. In this case use an approximate method to find the probability that during a period of one year (12 months) the shop will sell at least 170 items. (*WJEC*)

16 a The random variable X has a Poisson distribution and is such that $P(X = 2) = 3P(X = 4)$. Find, correct to three decimal places, the values of   (i) $P(X = 0)$,   (ii) $P(X \leqslant 4)$.

b The number of characters that are mistyped by a copy typist in any assignment has a Poisson distribution, the average number of mistyped characters per page being 0.8. In an assignment of 80 pages calculate, to three decimal places,

(i)   the probability that the first page will contain exactly two mistyped characters,

(ii) the probability that the first mistyped character will appear on the third page,

(iii) an approximate value for the probability that the total number of mistyped characters in the 80 pages will be at most 50.

(*WJEC*)

17 The number of organic particles suspended in a volume V cm^3 of a certain liquid follows a Poisson distribution with mean $0.1V$.

    a Find the probabilities that a sample of 1 cm^3 of the liquid will contain
       (i) at least one organic particle,
      (ii) exactly one organic particle.

    b Use an appropriate approximate procedure to find the probability that a sample of 1000 cm^3 of the liquid will contain at least 90 organic particles.

    c The liquid is sold in vials, each vial containing 10 cm^3 of the liquid. The vials are dispatched for sale in boxes, each box containing 100 vials. Find the probability that a vial will contain at least one organic particle. Hence find the mean and standard deviation of the number of vials per box of 100 vials that contain at least one organic particle.

                                                             (WJEC)

18 Records are kept daily of the number of times a certain fire brigade is called out. For a sample of 200 days the frequency distribution of the number of times per day that the brigade was called out is shown in the following table.

Number of calls	0	1	2	3	4	5	6	7
Number of days	31	53	51	33	20	8	2	2

   (i) Prepare a table for the given data showing all the terms in the summations that have to be calculated to obtain the mean and variance of this sample. Show that the mean is 2 and calculate the variance.

  (ii) It has been suggested that the number of times the fire brigade will be called out in a day has a Poisson distribution for which the probability of the brigade being called out k times in a day is

$$p(k) = e^{-m}\frac{m^k}{k!}, \qquad k = 0, 1, 2, \ldots,$$

       where m is the mean number of calls per day. List the sample relative frequencies and, taking $m = 2$, the corresponding Poisson probabilities. State the Poisson variance. Give reasons for believing that the Poisson distribution is appropriate.

Use the above Poisson distribution to obtain an estimate of the probability that on two randomly chosen days the fire brigade will be called out four times in all.   *(JMB)*

19 The number of emissions from a radioactive source in a period of t seconds has a Poisson distribution with mean $0.1t$.

   (i) Using the tables provided, or otherwise, calculate the probability that there will be at least three emissions in a period of one minute, giving your answer to three decimal places.

  (ii) Use a normal approximation to calculate the probability that no more than 25 emissions will occur in a period of five minutes, giving your answer to three decimal places.

(iii) The probability that at least one emission will occur in a period of b seconds is 0.99. Find the value of b, correct to the nearest integer.

(iv) The probability that no more than one emission will occur in a period of c seconds is 0.7. Show that c satisfies the equation

$$10 + c - 7e^{0.1c} = 0.$$

Verify that c lies in the interval $8 \leqslant c \leqslant 12$. Hence, using the method of interval bisection, or otherwise, find the value of c, correct to the nearest integer. *(JMB)*

20 The number of goals X scored by football team A in home matches against team B has a Poisson distribution with mean 2. Verify that

$$rP(X = r) = 2P(X = r - 1), \qquad r = 1, 2, \ldots$$

The number of goals Y scored by team B in away matches against team A has a Poisson distribution with mean 1. Write the relationship connecting $P(Y = r)$ and $P(Y = r - 1)$, for $r > 1$. Assuming that X and Y are independent, find the probabilities, correct to two decimal places, that in a match between A and B, on A's home ground,

(i) neither A nor B scores,

(ii) A and B score an equal, non-zero number of goals,

(iii) A scores more goals than B. *(JMB)*

Chapter 2

The distribution of a function of a random variable

The properties of the probability density function and the (cumulative) distribution function of a continuous random variable were studied in Chapter 8 of Book 1. Some of those properties are stated below.

If X is a continuous random variable with probability density function f, distribution function F and range space $\{x: a \leqslant x \leqslant b\}$ then, by definition,

$$F(x) = P(X \leqslant x),$$

therefore

$$F(x) = \int_a^x f(t)dt, \qquad a \leqslant x \leqslant b,$$

and since integration is the inverse operation to differentiation

$$f(x) = F'(x).$$

The above results are useful in determining the probability density function of a random variable Y which is a function h of another random variable X whose distribution is known. Initially, only problems in which h is a 1-1 function for values in the range space of X, will be considered.

2.1 Examples

In all the examples which follow, g and G will denote respectively the probability density function and the cumulative distribution function of Y.

Example 1
The random variable X has a probability density function f given by
$$f(x) = 3x^2, \qquad 0 \leqslant x \leqslant 1,$$
$$f(x) = 0, \qquad \text{otherwise.}$$
Find the probability density function of Y, where $Y = \dfrac{(1-X)}{2}$.

The method starts by trying to find an expression for $G(y)$.
By definition
$$G(y) = P(Y \leqslant y)$$

Substitute for Y $\qquad G(y) = P\left(\dfrac{1-X}{2} \leqslant y\right)$

Rearrange to make X the subject $\qquad G(y) = P(1 - 2y \leqslant X)$

This may be written $\qquad G(y) = P(X \geqslant 1 - 2y)$

Since $P(X \geqslant x) = 1 - P(X \leqslant x)$ $\qquad G(y) = 1 - P(X \leqslant 1 - 2y)$

Since $P(X \leqslant x) = F(x)$ $\qquad G(y) = 1 - F(1 - 2y)$ $\qquad$ (1)

Differentiate both sides with respect to y, using the chain rule on the right hand side.

$$G'(y) = -F'(1 - 2y).(-2)$$

But $G' = g$ and $F' = f$ $\qquad g(y) = +2f(1 - 2y)$

Since $f(x) = 3x^2$ for $0 \leqslant x \leqslant 1$, $f(1 - 2y) = 3(1 - 2y)^2$ for $0 \leqslant 1 - 2y \leqslant 1$,

$$g(y) = 2.3(1 - 2y)^2, \, 0 \leqslant y \leqslant \frac{1}{2}.$$

Therefore the probability density function g of Y is given by

$$g(y) = 6(1 - 2y)^2, \, 0 \leqslant y \leqslant \frac{1}{2},$$
$$g(y) = 0, \qquad \text{otherwise.}$$

In the above example it was not necessary to find $F(x)$ and $G(y)$ explicitly in order to find $g(y)$. However, if $F(x)$ is known then $G(y)$ may be found explicitly on line (1) before proceeding with the differentiation. This variation of the method is illustrated in **Example 2**.

Example 2

The random variable X has a probability density function f given by

$$f(x) = \frac{x}{2}, \qquad 0 \leqslant x \leqslant 2,$$
$$f(x) = 0, \qquad \text{otherwise.}$$

a Find the distribution function F of X.

b Find the probability density function of Y, where $Y = X^2$.

a For $0 \leqslant x \leqslant 2$, F is given by $\qquad F(x) = \displaystyle\int_0^x \frac{x}{2}\,dx$

$$= \left[\frac{x^2}{4}\right]_0^x$$

$$= \frac{x^2}{4}$$

Thus the distribution function F is given by

$$F(x) = 0, \qquad\qquad x < 0,$$
$$F(x) = \frac{x^2}{4}, \qquad\qquad 0 \leqslant x \leqslant 2,$$
$$F(x) = 1, \qquad\qquad x > 2.$$

b Again the method starts by trying to find an expression for $G(y)$.

By definition	$G(y) = P(Y \leqslant y)$
Substitute for Y	$G(y) = P(X^2 \leqslant y)$
Rearrange to make X the subject	$G(y) = P(-y^{\frac{1}{2}} \leqslant X < y^{\frac{1}{2}})$
Since $P(X \leqslant x) = F(x)$	$G(y) = F(y^{\frac{1}{2}}) - F(-y^{\frac{1}{2}})$

Since $F(x) = \dfrac{x^2}{4}$ for $0 \leqslant x \leqslant 2$ and $F(x) = 0$ for $x < 0$, it follows that

$F(y^{\frac{1}{2}}) = \dfrac{(y^{\frac{1}{2}})^2}{4}$ and $F(-y^{\frac{1}{2}}) = 0$ for $0 \leqslant y^{\frac{1}{2}} \leqslant 2$.

Therefore
$$G(y) = \frac{y}{4}, \quad 0 \leqslant y \leqslant 4$$

Differentiating with respect to y
$$G'(y) = \frac{1}{4}, \quad 0 \leqslant y \leqslant 4$$

Since $G' = g$
$$g(y) = \frac{1}{4}, \quad 0 \leqslant y \leqslant 4$$

Therefore the probability density function g of Y is given by
$$g(y) = \frac{1}{4}, \quad 0 \leqslant y \leqslant 4,$$
$$g(y) = 0, \quad \text{otherwise.}$$

Alternative method

When Y is a 1-1 function h for values in the range space of X, a somewhat quicker way of obtaining the probability density function uses the formula

$$g(y) = f(x)\left|\frac{dx}{dy}\right|,$$

where it is implicit that x is expressed in terms of y on the right hand side. A proof of this formula in the case when h is a strictly increasing function is given below.

Proof

When h is a strictly increasing function the inverse function h^{-1} exists.

Thus $Y = h(X)$ implies that $X = h^{-1}(Y)$.

By definition	$G(y) = P(Y \leqslant y)$.
Substitute for Y	$G(y) = P(h(X) \leqslant y)$.

Since h is an increasing function, when X is made the subject of the inequation it follows that $G(y) = P(X \leqslant h^{-1}(y))$.

Since $P(X \leqslant x) = F(x)$ $\qquad G(y) = F(h^{-1}(y))$. $\hspace{2cm}$ (1)

Differentiate with respect to y, using the chain rule on the right hand side.

$$G'(y) = F'(h^{-1}(y)) \cdot \frac{d}{dy}(h^{-1}(y))$$

Since $F' = f$ and $G' = g$ $\qquad g(y) = f(h^{-1}(y)) \cdot \dfrac{d}{dy}(h^{-1}(y)).$

Since $h^{-1}(y) = x$ this may be written in the less precise but more easily remembered form

$$g(y) = f(x) \cdot \frac{dx}{dy}. \tag{2}$$

The proof of the formula in the case where h is a strictly decreasing function follows a similar pattern and is left as an exercise for the reader. It should be noted that in this case line (1) becomes

$$G(y) = 1 - F(h^{-1}(y))$$

and line (2) is replaced by

$$g(y) = -f(x) \cdot \frac{dx}{dy}.$$

The results in the two cases may be amalgamated in the formula

$$g(y) = f(x)\left|\frac{dx}{dy}\right|.$$

Alternative solution to Example 1

$Y = \dfrac{(1 - X)}{2}$ is a 1-1 function of X.

Since $y = \dfrac{(1 - x)}{2}$ it follows that $x = 1 - 2y$ and $\dfrac{dx}{dy} = -2$.

Using $g(y) = f(x)\left|\dfrac{dx}{dy}\right|$ $\qquad\qquad g(y) = f(1 - 2y).|-2|.$

Since $f(x) = 3x^2$ for $0 \leqslant x \leqslant 1$, $f(1 - 2y) = 3(1 - 2y)^2$ for $0 \leqslant 1 - 2y \leqslant 1$.

Therefore $\qquad\qquad\qquad\qquad g(y) = 3(1 - 2y)^2 \times 2, \quad 0 \leqslant y \leqslant \dfrac{1}{2}.$

The probability density function g of Y is given by

$$g(y) = 6(1 - 2y)^2, \quad 0 \leqslant y \leqslant \frac{1}{2}$$
$$g(y) = 0, \qquad\qquad\text{otherwise,}$$

as was shown earlier.

Alternative solution to Example 2b

$Y = X^2$ is a 1-1 function for values of X in its range space, viz. $0 \leqslant x \leqslant 2$.

Since $y = x^2$ it follows that $x = y^{\frac{1}{2}}$ and $\dfrac{dx}{dy} = \dfrac{1}{2}y^{-\frac{1}{2}}$.

Using $g(y) = f(x)\left|\dfrac{dx}{dy}\right|$ $\qquad\qquad g(y) = f(y^{\frac{1}{2}})\left|\dfrac{1}{2}y^{-\frac{1}{2}}\right|.$

Since $f(x) = \dfrac{x}{2}$ for $0 \leqslant x \leqslant 2$, $\qquad f(y^{\frac{1}{2}}) = \dfrac{y^{\frac{1}{2}}}{2}$ $\qquad$ for $0 \leqslant y^{\frac{1}{2}} \leqslant 2$

$$g(y) = \frac{y^{\frac{1}{2}}}{2} \cdot \frac{1}{2} y^{-\frac{1}{2}} \qquad \text{for } 0 \leqslant y \leqslant 4.$$

Therefore the probability density function g of Y is given by

$$g(y) = \frac{1}{4}, \quad 0 \leqslant y \leqslant 4,$$

$$g(y) = 0, \quad \text{otherwise.}$$

These solutions are shorter than the previous ones, but it must be clearly understood that the formula $g(y) = f(x)\left|\dfrac{dx}{dy}\right|$ may only be used when Y is a 1-1 function for all values in the range space of X.

Exercise 2.1

1 The probability density function f of a random variable X is given by
$$f(x) = \frac{1}{5}, \qquad\qquad 2 \leqslant x \leqslant 7,$$
$$f(x) = 0, \qquad\qquad \text{otherwise.}$$
Find the probability density function of Y, where $Y = 3X - 1$.

2 The probability density function f of a random variable X is given by
$$f(x) = \frac{(2 - x)}{2}, \qquad\qquad 0 \leqslant x \leqslant 2,$$
$$f(x) = 0, \qquad\qquad \text{otherwise.}$$
Find the probability density function of Y, where $Y = 2 - X$.

3 The probability density function f of a random variable X is given by
$$f(x) = x\frac{(4 - x)}{9}, \qquad\qquad 0 \leqslant x \leqslant 3,$$
$$f(x) = 0, \qquad\qquad \text{otherwise.}$$
Find the probability density function of Y, where $Y = X^2$.

4 The probability density function f of a random variable X is given by
$$f(x) = \frac{1}{4}, \qquad\qquad 4 \leqslant x \leqslant 8,$$
$$f(x) = 0, \qquad\qquad \text{otherwise.}$$
Find the probability density function of Y, where $Y = \dfrac{(16 - 3x)}{4}$.

5 The probability density function f of a random variable X is given by
$$f(x) = 2(2 - x), \qquad\qquad 1 \leqslant x \leqslant 2,$$
$$f(x) = 0, \qquad\qquad \text{otherwise.}$$
Find the probability density function of Y, where $Y = \dfrac{1}{X}$.

23

2.2 Some harder examples

The examples and the exercises above were relatively straightforward as, in each case, Y was an easy 1-1 function of X; in the following examples this is not so.

Example 3
The probability density function of a random variable X is given by
$$f(x) = \frac{x}{4}, \qquad 1 \leqslant x \leqslant 3,$$
$$f(x) = 0, \qquad \text{otherwise.}$$
Find the probability density function of Y, where $Y = |X - 2|$.

In this question Y is not a 1-1 function of X for values in the range space of X, as may be seen from its graph, shown below. The alternative method, using $g(y) = f(x)\left|\dfrac{dx}{dy}\right|$ should not be used.

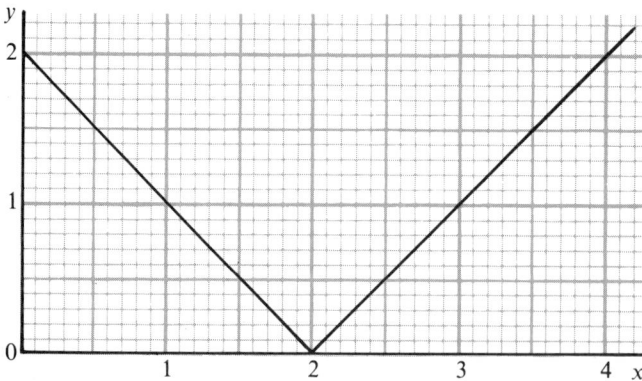

Graph of $y = |x - 2|$

By definition	$G(y) = P(Y \leqslant y)$.
Substitute for Y	$G(y) = P(|X - 2| \leqslant y)$.
Therefore	$G(y) = P(-y \leqslant X - 2 \leqslant y)$.
Rearranging	$G(y) = P(2 - y \leqslant X \leqslant 2 + y)$
	$= P(Y \leqslant 2 + y) - P(Y \leqslant 2 - y)$.
Therefore	$G(y) = F(2 + y) - F(2 - y)$.

Differentiating with respect to y, using the chain rule on the right hand side
$$G'(y) = F'(2 + y) - F'(2 - y).(-1).$$
But $G' = g$ and $F' = f$ $\qquad g(y) = f(2 + y) + f(2 - y)$. $\qquad$ (1)

As X takes values from 1 to 2 to 3, Y takes values from 1 to 0 to 1; and since $f(x) = \dfrac{x}{4}$ for $1 \leqslant x \leqslant 3$, it follows that $f(2 + y) = \dfrac{(2 + y)}{4}$ and $f(2 - y) = \dfrac{(2 - y)}{4}$ for $0 \leqslant y \leqslant 1$.

Substituting in (1) $g(y) = \frac{1}{4}(2 + y) + \frac{1}{4}(2 - y)$ for $0 \leqslant y \leqslant 1$

$$\therefore \quad g(y) = 1.$$

Therefore the probability density function g of Y is given by

$$g(y) = 1, \qquad 0 \leqslant y \leqslant 1,$$
$$g(y) = 0, \qquad \text{otherwise.}$$

Example 4

The probability density function f of a random variable X is given by

$$f(x) = \frac{3x^2}{5}, \qquad -1 \leqslant x \leqslant 1,$$

$$f(x) = \frac{3}{5}, \qquad 1 < x \leqslant 2,$$

$$f(x) = 0, \qquad \text{otherwise.}$$

Find the probability density function of Y, where $Y = X^2$.

In this question Y is not a 1-1 function of X for values of X between -1 and 1.

By definition $G(y) = P(Y \leqslant y)$.

Substitute for Y $G(y) = P(X^2 \leqslant y)$.

Therefore $G(y) = P(-y^{\frac{1}{2}} \leqslant X \leqslant y^{\frac{1}{2}})$

$$G(y) = P(X \leqslant y^{\frac{1}{2}}) - P(X \leqslant -y^{\frac{1}{2}})$$
$$G(y) = F(y^{\frac{1}{2}}) - F(-y^{\frac{1}{2}}).$$

Differentiating with respect to y, using the chain rule on the right hand side

$$G'(y) = F'(y^{\frac{1}{2}}) \cdot \frac{1}{2} y^{-\frac{1}{2}} - F'(-y^{\frac{1}{2}}) \cdot \left(-\frac{1}{2}\right) y^{-\frac{1}{2}}.$$

Since $G' = g$ and $F' = f$ $g(y) = f(y^{\frac{1}{2}}) \cdot \frac{1}{2} y^{-\frac{1}{2}} + f(-y^{\frac{1}{2}}) \cdot \frac{1}{2} y^{-\frac{1}{2}}$ (1)

Since f is a 'split' function it is necessary to give separate consideration to $-1 \leqslant x \leqslant 1$ and $1 < x \leqslant 2$.

As X takes values from -1 to 0 to 1, Y takes values from 1 to 0 to 1; since $f(x) = \frac{3x^2}{5}$ for $-1 \leqslant x \leqslant 1$ it follows that $f(y^{\frac{1}{2}}) = \frac{3y}{5}$ and $f(-y^{\frac{1}{2}}) = \frac{3y}{5}$ for $0 \leqslant y \leqslant 1$.

Substituting in (1) $g(y) = \frac{3}{5} y \cdot \frac{1}{2} y^{-\frac{1}{2}} + \frac{3}{5} y \cdot \frac{1}{2} y^{-\frac{1}{2}}$ $0 \leqslant y \leqslant 1$.

Therefore $g(y) = \frac{3}{5} y^{\frac{1}{2}}$ $0 \leqslant y \leqslant 1$.

As X takes values from 1 to 2, Y takes values from 1 to 4; since $f(x) = \frac{3}{5}$ for $1 < x \leqslant 2$ and $f(x) = 0$ for $x < -1$ it follows that $f(y^{\frac{1}{2}}) = \frac{3}{5}$ and $f(-y^{\frac{1}{2}}) = 0$ for $1 < y \leqslant 4$.

25

Substituting in (1) $\qquad g(y) = \dfrac{3}{5} \cdot \dfrac{1}{2} y^{-\frac{1}{2}} + 0, \qquad\qquad 1 < y \leqslant 4.$

Therefore $\qquad\qquad\quad g(y) = \dfrac{3}{10} y^{-\frac{1}{2}}, \qquad\qquad\qquad 1 < y \leqslant 4.$

Therefore the probability density function g of Y is given by

$$g(y) = \frac{3}{5} y^{\frac{1}{2}}, \qquad\qquad 0 \leqslant y \leqslant 1,$$

$$g(y) = \frac{3}{10} y^{-\frac{1}{2}}, \qquad\qquad 1 < y \leqslant 4,$$

$$g(y) = 0, \qquad\qquad\qquad \text{otherwise.}$$

Example 5

A random variable X is uniformly distributed between 2 and 5. Find the distribution function of X.

Let Y denote the larger of two random observations of X. Find the probability density function of Y.

Since $X \sim U(2, 5)$ the probability density function f of X is given by

$$f(x) = \frac{1}{3}, \qquad\qquad 2 \leqslant x \leqslant 5$$

$$F(x) = \int_2^x \frac{1}{3} dx \qquad\qquad 2 \leqslant x \leqslant 5$$

$$= \left[\frac{1}{3} x \right]_2^x$$

$$= \frac{1}{3}(x - 2).$$

The distribution function F of X is given by

$$F(x) = 0 \qquad\qquad\qquad\qquad x < 2$$

$$F(x) = \frac{1}{3}(x - 2) \qquad\qquad\quad 2 \leqslant x \leqslant 5$$

$$F(x) = 1 \qquad\qquad\qquad\qquad x > 5.$$

By definition $\qquad G(y) = P(Y \leqslant y)$

$$= P(\text{max. of } X_1, X_2 \leqslant y)$$

$$= P(X_1 \leqslant y \text{ and } X_2 \leqslant y).$$

Since X_1, X_2 are independent

$$G(y) = P(X_1 \leqslant y) \cdot P(X_2 \leqslant y).$$

Since X_1, X_2 each have the same distribution as X

$$G(y) = F(y) \cdot F(y)$$

$$= \frac{1}{3}(y - 2) \cdot \frac{1}{3}(y - 2) \text{ for } 2 \leqslant y \leqslant 5$$

$$= \frac{1}{9}(y - 2)^2.$$

Differentiate with respect to y and using $G' = g$

$$g(y) = \frac{2}{9}(y - 2).$$

Therefore the probability density function of Y is given by

$$g(y) = \frac{2}{9}(y - 2), \qquad 2 \leqslant y \leqslant 5,$$
$$g(y) = 0, \qquad \text{otherwise.}$$

Exercise 2.2

1 The probability density function f of the random variable X is given by

$$f(x) = \frac{6x^2}{11}, \qquad 0 \leqslant x \leqslant 1,$$
$$f(x) = \frac{6x}{11}, \qquad 1 < x \leqslant 2,$$
$$f(x) = 0, \qquad \text{otherwise.}$$

Find the probability density function of Y, where $Y = \frac{(X - 1)}{3}$.

2 The probability density function f of the random variable X is given by

$$f(x) = \frac{1}{x^2}, \qquad 1 \leqslant x \leqslant 3,$$
$$f(x) = \frac{1}{9}, \qquad 3 < x \leqslant 6,$$
$$f(x) = 0, \qquad \text{otherwise.}$$

Find the probability density function of Y, where $Y = \frac{1}{X^2}$.

3 The probability density function f of the random variable X is given by

$$f(x) = \frac{x}{2}, \qquad 0 \leqslant x \leqslant 2,$$
$$f(x) = 0, \qquad \text{otherwise.}$$

Find the probability density function of Y, where $Y = |X - 1|$.

4 The probability density function f of the random variable X is given by

$$f(x) = x\frac{(4 - x)}{9}, \qquad 0 \leqslant x \leqslant 3,$$
$$f(x) = 0, \qquad \text{otherwise.}$$

Find the probability density function of Y, where $Y = |X - 1|$.

5 The probability density function f of the random variable X is given by

$$f(x) = \frac{1}{3}, \qquad 2 \leqslant x \leqslant 5,$$
$$f(x) = 0, \qquad \text{otherwise.}$$

Find the probability density function of Y, where Y is the smaller of two randomly selected values of X.

Miscellaneous Exercise ☐ 2

1 A machine produces square plates. The lengths of the sides of the squares are uniformly distributed between 9.9 cm and 10.1 cm. The area of a square plate is denoted by Y cm^2. Find the probability density function of Y. (*JMB*)

2 Derive the mean and variance of the continuous random variable X having probability density function

$$f(x) = (b - a)^{-1}, \qquad a \leqslant x \leqslant b,$$
$$f(x) = 0, \qquad \text{otherwise.}$$

Obtain the probability density function of

$$Y = \frac{X - a}{b - a},$$

stating the range of values of y for which the density function is non-zero. Given that $Z = \ln Y$, show that the probability density function of Z is

$$g(z) = e^z, \qquad z \leqslant 0,$$
$$g(z) = 0, \qquad z > 0.$$

Find the probability that a randomly chosen value, z, of Z lies in the interval $-3 \leqslant z \leqslant 0$.

Find the mean and variance of Z.

$$\left[\text{You may assume that } \int_{-\infty}^{0} z^n e^z \, dz = (-1)^n n! \right.$$

for every non-negative integer n. $\Big]$

(*JMB*)

3 The average speed X, in m.p.h., of a bus on a certain journey of 5 miles is a continuous random variable with a probability density function given by

$$f(x) = \frac{k}{x^2}, \qquad 20 \leqslant x \leqslant 30,$$
$$f(x) = 0, \qquad \text{otherwise.}$$

Show that $k = 60$.

Find (i) the mean of X,
(ii) the distribution function of X for all values of x,
(iii) the median value of X,
(iv) the probability density function of the time, Y minutes, which the bus takes on the journey. (*JMB*)

4 AB is the diameter of a semi-circle with radius r metres and P is a point chosen at random on the semi-circle. The size of angle ABP is X radians, where X is uniformly distributed between 0 and $\frac{\pi}{2}$. Write down the probability density function and the distribution function of X. The length of the chord AP is Y metres. Find the probability density function of Y. (*JMB*)

5 The random variable X is rectangularly distributed in the interval $0 \leqslant x \leqslant \pi$. The random variable Y is a transformation of X given by

$$y = \tan\left(\frac{x}{2} - \frac{\pi}{4}\right).$$

Show that the probability density function, $g(y)$, of Y is given by

$$g(y) = \frac{2}{\pi(1 + y^2)}, \qquad a \leqslant y \leqslant b,$$

$$g(y) = 0, \qquad\qquad \text{otherwise},$$

and state the values of a and b.

Sketch the graph of $g(y)$ and state the mean value of Y. $\hspace{2em}(JMB)$

6 The maximum length to which a string of natural length a metres can be stretched before it snaps is $a(1 + X^2)$ metres, where X is a continuous random variable whose probability density function is

$$f(x) = 4x, \qquad 0.25 \leqslant x \leqslant 0.75,$$

$$f(x) = 0, \qquad \text{otherwise}.$$

(i) Calculate the probability that a string can be stretched to one and a half times its natural length without snapping.

(ii) Find the value of $E[X^2]$ and hence find μ, the mean maximum stretched length of strings of natural length 1 metre.

(iii) Find the probability density function of $Y = 1 + X^2$, the maximum stretched length of a string of natural length 1 metre, and use it to verify the value of μ you obtained in (ii). $\hspace{2em}(WJEC)$

7 The continuous random variable X has probability density function f, where

$$f(x) = cx(2 - x), \qquad 0 \leqslant x \leqslant 2,$$

$$f(x) = 0, \qquad\qquad \text{otherwise}.$$

(i) Find the value of c.

(ii) Show that the variance of X is 0.2.

(iii) Determine the cumulative distribution function of X.

(iv) Given that $\qquad y = \dfrac{x}{4 - x}$

determine the range R of the values of y corresponding to values of x ranging from 0 to 2. Show that the probability density function g of

$$Y = \frac{X}{4 - X}$$

is given by $\qquad g(y) = \dfrac{ky(1 - y)}{(1 + y)^4} \qquad$ for y in R.

$$g(y) = 0, \qquad\qquad \text{otherwise},$$

and write down the value of the constant k. $\hspace{2em}(WJEC)$

8 The continuous random variable X has probability density function f, where

$$f(x) = \frac{1}{2}, \qquad\qquad 0 \leqslant x < 1,$$

$$f(x) = \frac{(3-x)}{4}, \qquad\qquad 1 \leqslant x \leqslant 3,$$

$$f(x) = 0, \qquad\qquad \text{otherwise.}$$

(i) Find the mean of the distribution.

(ii) Obtain expressions for $F(x)$, where F is the cumulative distribution function of X.

(iii) Find the value of a given that $P(X \leqslant a) = \frac{7}{9}$.

(iv) Determine the probability density function of $Y = 3 - X$. *(WJEC)*

9 The continuous random variable X has probability density function f, where

$$f(x) = \frac{25}{12(x+1)^3}, \qquad\qquad 0 \leqslant x \leqslant 4,$$

$$f(x) = 0, \qquad\qquad \text{otherwise.}$$

(i) Evaluate $E[X + 1]$. Hence, or otherwise, find the mean of X.

(ii) Find the value for $c > 0$ for which $P(X \leqslant c) = c$.

(iii) Find the cumulative distribution function of $Y = (X + 1)^{-2}$. Hence, or otherwise, find the probability density function of Y. *(WJEC)*

10 The continuous random variable X has probability density function f, where

$$f(x) = \frac{1}{8}, \qquad\qquad \text{for } 0 \leqslant x < 2,$$

$$f(x) = \frac{x}{8}, \qquad\qquad \text{for } 2 \leqslant x \leqslant 4,$$

$$f(x) = 0, \qquad\qquad \text{otherwise.}$$

(i) Obtain expressions for $F(x)$, where F is the cumulative distribution function of X.

(ii) Evaluate (a) $P(1 \leqslant X \leqslant 3)$,   (b) $P(X \geqslant 1 | X \leqslant 3)$.

(iii) Let g denote the probability density function of $Y = |X - 3|$. Show that

$$g(y) = \frac{1}{8}, \qquad\qquad \text{for } 1 < y \leqslant 3.$$

Find $g(y)$ for all other values of y for which $g(y)$ is non-zero. Hence, or otherwise, evaluate $E[|X - 3|]$. *(WJEC)*

11 Manufactured steel rods have lengths that are rectangularly distributed over the interval from 2 to 3, inclusive. Two of these rods are chosen at random. Let U denote the length of the longer of the two rods. Find an expression for $P(U \leqslant u)$, where $2 \leqslant u \leqslant 3$, and hence, or otherwise, show that the probability density function of U is given by

$$g(u) = 2(u - 2), \qquad\qquad 2 \leqslant u \leqslant 3,$$
$$g(u) = 0, \qquad\qquad \text{otherwise.}$$

Also find the probability density function of the length, W, of the shorter of the two rods.

Two further rods are chosen at random. Find the probability density function of the length of the longest of all four rods. Show that the mean lengths of the longest and the shortest of the four rods are in the ratio 14:11.

(JMB)

12 The continuous random variable X has probability density function f, where

$$f(x) = \frac{2}{5}, \qquad\qquad 0 < x < 1,$$

$$f(x) = \frac{2}{15}(4 - x), \qquad\qquad 1 \leqslant x \leqslant 4,$$

$$f(x) = 0, \qquad\qquad \text{otherwise.}$$

(i) Find the distribution function F of X. Hence, or otherwise, determine the value of the lower quartile of X and, correct to two decimal places, the value of the median of X.

(ii) Let G denote the distribution function of the random variable Y given by

$$Y = \frac{12}{X + 2}.$$

Show that, for $2 \leqslant y \leqslant 4$,

$$G(y) = \frac{12}{5}\left(1 - \frac{2}{y}\right)^2.$$

Obtain an expression for $G(y)$ for $4 < y < 6$.

Hence, or otherwise, obtain expressions for $g(y)$, where g is the probability density function of Y.

(JMB)

Chapter 3

Joint distributions of discrete random variables

The distributions studied so far have been distributions of a single random variable, but there are many experiments in which there are two random variables associated with the outcomes of the experiment. For example, the height and weight of a randomly chosen person may be measured or the number of children and the number of bedrooms in the home of a randomly chosen family may be recorded. Initially only joint distributions of two discrete random variables will be studied; the *joint distribution* of two discrete random variables X and Y, either associated with the outcomes of a single random experiment or with the outcomes of two separate random experiments, consists of the specification of the values of $P(X = x_i \cap Y = y_j)$ for all possible pairs of values of x_i and y_j.

3.1 Introductory examples

Example 1

A fair cubical die has four faces numbered 0 and the other two faces numbered 1. The die is thrown three times. Let X be the total score on the three throws and let Y be the score on the first two throws. Find the joint probability distribution of X and Y.

Consider each possible ordered outcome, calculate the values of X and Y associated with the outcome and the probability of the outcome.

Outcome	X	Y	Probability	
$(0, 0, 0)$	0	0	$\frac{2}{3} \times \frac{2}{3} \times \frac{2}{3} = \frac{8}{27}$	$P(X = 0 \cap Y = 0) = 8/27$
$(0, 0, 1)$	1	0	$\frac{2}{3} \times \frac{2}{3} \times \frac{1}{3} = \frac{4}{27}$	$P(X = 1 \cap Y = 0) = 4/27$
$(0, 1, 0)$	1	1	$\frac{2}{3} \times \frac{1}{3} \times \frac{2}{3} = \frac{4}{27}$	
$(1, 0, 0)$	1	1	$\frac{1}{3} \times \frac{2}{3} \times \frac{2}{3} = \frac{4}{27}$	$P(X = 1 \cap Y = 1) = 8/27$

$(0, 1, 1)$ 2   1 $\dfrac{2}{3} \times \dfrac{1}{3} \times \dfrac{1}{3} = \dfrac{2}{27}$

$(1, 0, 1)$ 2   1 $\dfrac{1}{3} \times \dfrac{2}{3} \times \dfrac{1}{3} = \dfrac{2}{27}$

$\left.\right\} \; P(X = 2 \cap Y = 1) = 4/27$

$(1, 1, 0)$ 2   2 $\dfrac{1}{3} \times \dfrac{1}{3} \times \dfrac{2}{3} = \dfrac{2}{27}$ $P(X = 2 \cap Y = 2) = 2/27$

$(1, 1, 1)$ 3   2 $\dfrac{1}{3} \times \dfrac{1}{3} \times \dfrac{1}{3} = \dfrac{1}{27}$ $P(X = 3 \cap Y = 2) = 1/27$

The joint distribution of X and Y may be displayed in a two-way table as shown below.

		\(x \)			
		0	1	2	3
	0	8/27	4/27	0	0
y	1	0	8/27	4/27	0
	2	0	0	2/27	1/2

Example 2

Find the distributions of X and Y, the random variables described in **Example 1** and deduce $E[X]$ and $E[Y]$.

The joint distribution table is recorded below together with an extra column headed $P(Y = y)$ and an extra row giving $P(X = x)$.

		\(x \)				
		0	1	2	3	$P(Y = y)$
	0	8/27	4/27	0	0	12/27
y	1	0	8/27	4/27	0	12/27
	2	0	0	2/27	1/27	3/27
$P(X = x)$		8/27	12/27	6/27	1/27	

Since $P(X = 0) = P(X = 0 \cap Y = 0) + P(X = 0 \cap Y = 1) + P(X = 0 \cap Y = 2)$, the first entry in the last row of the above table is found by adding the probabilities in the first column of the joint probability table. The second entry in the last row is found by adding the probabilities in the second column, etc.

Similarly the entries in the last column of the above table are found by adding the probabilities in the corresponding rows of the joint probability table.

The distribution of X is given by the first and last rows of the above table whilst the distribution of Y is given by the first and last columns.

The distributions of X and Y deduced from the joint distribution in this manner are often referred to as the marginal distributions of X and Y, although the use of the adjective "marginal" is superfluous.

<table>
<tr><th colspan="3">Distribution of X</th><th colspan="3">Distribution of Y</th></tr>
<tr><th>x_i</th><th>p_i</th><th>$p_i x_i$</th><th>y_i</th><th>p_i</th><th>$p_i y_i$</th></tr>
<tr><td>0</td><td>8/27</td><td>0</td><td>0</td><td>12/27</td><td>0</td></tr>
<tr><td>1</td><td>12/27</td><td>12/27</td><td>1</td><td>12/27</td><td>12/27</td></tr>
<tr><td>2</td><td>6/27</td><td>12/27</td><td>2</td><td>3/27</td><td>6/27</td></tr>
<tr><td>3</td><td>1/27</td><td>3/27</td><td></td><td>27/27</td><td>18/27</td></tr>
<tr><td></td><td>27/27</td><td>27/27</td><td></td><td></td><td></td></tr>
</table>

$$E[X] = 27/27 = 1 \qquad\qquad E[Y] = 18/27 = 2/3$$

Example 3

Given that X and Y are the random variables described in **Example 1** and that $U = X + Y$ and $V = XY$, find

a the distribution of U and $E[U]$,

b the distribution of V and $E[V]$.

a In the following joint distribution table of X and Y, the value of U for each pair of values of x, y has been placed in brackets after the associated probability.

			x		
		0	1	2	3
	0	8/27(0)	4/27(1)	0	0
y	1	0	8/27(2)	4/27(3)	0
	2	0	0	2/27(4)	1/27(5)

The distribution of U is given by the first two columns of the table below.

u_i	p_i	$p_i u_i$
0	8/27	0
1	4/27	4/27
2	8/27	16/27
3	4/27	12/27
4	2/27	8/27
5	1/27	5/27
	27/27	45/27

$$E[U] = 45/27 = 5/3$$

It may be noted that $E[X + Y] = E[X] + E[Y]$.

b In the joint distribution table below, the value of V for each pair of values x, y has been placed in brackets after the associated probability.

	x 0	1	2	3
0	8/27(0)	4/27(0)	0	0
y 1	0	8/27(1)	4/27(2)	0
2	0	0	2/27(4)	1/27(6)

The distribution of V is given by the first two columns of the table below.

v_i	p_i	$p_i v_i$
0	12/27	0
1	8/27	8/27
2	4/27	8/27
4	2/27	8/27
6	1/27	6/27
	27/27	30/27

$$E[V] = 30/27 = 10/9$$

It may be noted that $E[XY] \neq E[X].E[Y]$.

If **Examples 1, 2** and **3** formed a single question, it would not be necessary to write out the joint distribution table four times as above, once would suffice but then the greatest care would be needed.

Exercise 3.1

1 Two balls are drawn at random without replacement from a bag which contains four white balls, three blue balls and two red balls. Let X denote the number of white balls drawn and let Y be the number of blue drawn.
   a Find the joint distribution of X and Y.
   b Find $E[X]$, $V[X]$, $E[Y]$, $V[Y]$, $E[X + Y]$ and $V[X + Y]$.
   c Show that $E[X + Y] = E[X] + E[Y]$.
   d Show that $V[X + Y] \neq V[X] + V[Y]$.

2 A fair coin is tossed three times in succession. Let X denote the total number of heads obtained and let Y denote the number of heads in the longest run of successive heads obtained.
   a Find the joint distribution of X and Y.
   b Find $E[X]$, $E[Y]$ and $E[XY]$.
   c Show that $E[XY] \neq E[X].E[Y]$.

3 An unbiased cubical die has three faces numbered 0, two faces numbered 1 and one face numbered 2. The die is thrown three times in succession. Let X denote the total score obtained and let Y denote the number of 0s obtained.

    a Find the joint distribution of X and Y.

    b Find $E[X]$, $E[Y]$ and $E[X + Y]$.

    c Show that $E[X + Y] = E[X] + E[Y]$.

3.2 Joint distribution of two discrete random variables

The ideas introduced in **Examples 1, 2** and **3** may be expressed more formally as follows.

Suppose that X is a discrete random variable with range space $\{x_1, x_2, \ldots, x_n\}$ and Y is another discrete random variable with range space $\{y_1, y_2, \ldots, y_m\}$.

Consider the two-dimensional random variable (X, Y) having a range space

$$R = \{(x_i, y_j): i = 1, 2, \ldots, n; \quad j = 1, 2, \ldots, m\}.$$

Let p be the function with domain R such that

$$p(x_i, y_j) = P(X = x_i \cap Y = y_j) \qquad \text{for all } i, j;$$

$$p(x_i, y_j) \geqslant 0 \qquad \text{for all } i, j;$$

$$\sum_{i=1}^{n} \sum_{j=1}^{m} p(x_i, y_j) = 1.$$

The function p is called the *joint probability function* of X and Y and a table giving the values of $p(x_i, y_j)$ for all pairs of values (x_i, y_j) is called the *joint distribution* of X and Y.

The following abbreviations will be used in the remainder of the chapter.

$$p_{ij} = P(X = x_i \cap Y = y_j),$$

$$p_i = P(X = x_i),$$

$$p'_j = P(Y = y_j).$$

Since the event $X = x_i$ must occur with one of the events $Y = y_1$, $Y = y_2$, $\ldots$, $Y = y_m$.

$$P(X = x_i) = P(X = x_i \cap Y = y_1) + P(X = x_i \cap Y = y_2) + \cdots + P(X = x_i \cap Y = y_m)$$

or using the above abbreviations

$$p_i = \sum_{j=1}^{m} p_{ij}$$

Similarly $\quad p'_j = \displaystyle\sum_{i=1}^{n} p_{ij}.$

Expected value

The *expected value* (or *expectation*) of a function $f(X, Y)$ is defined as

$$E[f(X, Y)] = \sum_{i=1}^{n} \sum_{j=1}^{m} p_{ij} f(x_i, y_j)$$

Using this definition

$$E[X] = \sum_{i=1}^{n} \sum_{j=1}^{m} p_{ij} x_i$$

$$= \sum_{i=1}^{n} (p_{i1} x_i + p_{i2} x_i + \cdots + p_{im} x_i)$$

$$= \sum_{i=1}^{n} x_i(p_{i1} + p_{i2} + \cdots + p_{im})$$

$$= \sum_{i=1}^{n} x_i \left(\sum_{j=1}^{m} p_{ij} \right)$$

$$= \sum_{i=1}^{n} x_i p_i.$$

This shows that the new definition is equivalent to the one for $E[X]$ given previously on page 97 of Book 1.

A useful result involving expected values is as follows:

If $f(X, Y) = ag(X, Y) + bh(X, Y)$, where a and b are constants,

then $\qquad E[f(X, Y)] = aE[g(X, Y)] + bE[h(X, Y)].$ \hfill (1)

Proof: $\qquad E[f(X, Y)] = \sum_{i=1}^{n} \sum_{j=1}^{m} p_{ij}\{ag(x_i, y_j) + bh(x_i, y_j)\}$

$$= \sum_{i=1}^{n} \sum_{j=1}^{m} p_{ij} ag(x_i, y_j) + \sum_{i=1}^{n} \sum_{j=1}^{m} p_{ij} bh(x_i, y_j)$$

$$= a \sum_{i=1}^{n} \sum_{j=1}^{m} p_{ij} g(x_i, y_j) + b \sum_{i=1}^{n} \sum_{j=1}^{m} p_{ij} h(x_i, y_j)$$

$$= aE[g(X, Y)] + bE[h(X, Y)].$$

Some examples of result (1) are given below

$$E[2X^2 + 5Y] \quad = 2E[X^2] + 5E[Y]$$
$$E[X(X + 2Y)] = E[X^2] + 2E[XY]$$
$$E[aX + bY] \quad = aE[X] + bE[Y].$$

The last of these is frequently used.

Independence of two discrete random variables

The discrete random variables X and Y are said to be *independent* if and only if

$$P(X = x_i \cap Y = y_j) = P(X = x_i). P(Y = y_j) \quad \text{or} \quad p_{ij} = p_i.p_j'$$

for *every* pair (x_i, y_j) in the range space R.

Example 4

The joint distribution of two discrete random variables X and Y is given in the following table in which p is a constant.

		x		
		0	1	2
y	0	p	$2p$	$3p$
	1	0	$4p$	$5p$

a Find the value of p.

b Show that X and Y are not independent.

c Show that $V[X + Y] \neq V[X] + V[Y]$.

In the table below the marginal distributions of X and Y are shown and the value of $(X + Y)$ for each pair (x, y) is given in a bracket following the associated probability.

		x			$P(Y = y)$
		0	1	2	
y	0	$p(0)$	$2p(1)$	$3p(2)$	$6p$
	1	$0(1)$	$4p(2)$	$5p(3)$	$9p$
$P(X = x)$		p	$6p$	$8p$	$15p$

a The sum of all the probabilities in the table is equal to 1. Therefore $15p = 1$ and $p = 1/15$.

b From the joint distribution table: $P(X = 0 \cap Y = 1) = 0$.

From the marginal distribution of X: $P(X = 0) = 1/15$.

From the marginal distribution of Y: $P(Y = 1) = 9/15$.

Since $P(X = 0 \cap Y = 1) \neq P(X = 0) . P(Y = 1)$ it follows that X and Y are not independent.

c

Distribution of X

x_i	p_i	$p_i x_i$	$p_i x_i^2$
0	1/15	0	0
1	6/15	6/15	6/15
2	8/15	16/15	32/15
	15/15	22/15	38/15

$E[X] = 22/15$

$V[X] = E[X^2] - E[X]^2$
$= 38/15 - (22/15)^2$
$= 86/225$

Distribution of Y

y_i	p_i	$p_i y_i$	$p_i y_i^2$
0	6/15	0	0
1	9/15	9/15	9/15
	15/15	9/15	9/15

$E[Y] = 9/15$

$V[Y] = E[Y^2] - E[Y]^2$
$= 9/15 - (9/15)^2$
$= 6/25$

Distribution of $Z = X + Y$

z_i	p_i	$p_i z_i$	$p_i z_i^2$
0	1/15	0	0
1	2/15	2/15	2/15
2	7/15	14/15	28/15
3	5/15	15/15	45/15
	15/15	31/15	75/15

$$E[Z] = 31/15$$
$$V[Z] = E[Z^2] - E[Z]^2$$
$$= 75/15 - (31/15)^2$$
$$= 164/225$$

$$V[X] + V[Y] = 86/225 + 6/25$$
$$= 140/225$$

$$V[X] + V[Y] \neq V[X + Y]$$

Example 5

The joint distribution of the two discrete random variables X and Y is given in the following table where p is a constant.

			x	
		1		2
y	0	p		$3p$
	1	$2p$		$6p$

a Find the value of p.

b Show that X and Y are independent.

c Find the distribution of $X - Y$ and verify that
$V[X - Y] = V[X] + V[Y]$.

In the table below the marginal distributions of X and Y are shown and the value of $X - Y$ for each pair (x, y) is given in a bracket following the associated probability.

		x		
		1	2	$P(Y = y)$
y	0	$p(1)$	$3p(2)$	$4p$
	1	$2p(0)$	$6p(1)$	$8p$
$P(X = x)$		$3p$	$9p$	$12p$

STATISTICS 2

a The sum of all the probabilities in the joint distribution table is 1.
Therefore $12p = 1$ and $p = 1/12$.

b From the joint distribution table: $\quad P(X = 1 \cap Y = 0) = 1/12.$
From the marginal distribution of X: $\quad P(X = 1) = 3/12 = 1/4.$
From the marginal distribution of Y: $\quad P(Y = 0) = 4/12 = 1/3.$
Therefore $\quad P(X = 1 \cap Y = 0) = P(X = 1) . P(Y = 0).$

Similarly
$$P(X = 2 \cap Y = 0) = 1/4 = 3/4 . 1/3 = P(X = 2) . P(Y = 0)$$
$$P(X = 1 \cap Y = 1) = 1/6 = 1/4 . 2/3 = P(X = 1) . P(Y = 1)$$
$$P(X = 2 \cap Y = 1) = 1/2 = 3/4 . 2/3 = P(X = 2) . P(Y = 1).$$

Therefore $\quad P(X = x \cap Y = y) = P(X = x) . P(Y = y)$ for every pair (x, y)
and X and Y are independent.

c

Distribution of X			
x_i	p_i	$p_i x_i$	$p_i x_i^2$
1	1/4	1/4	1/4
2	3/4	6/4	12/4
	4/4	7/4	13/4

$E[X] = 7/4$

$$V[X] = E[X^2] - E[X]^2$$
$$= 13/4 - (7/4)^2$$
$$= 3/16$$

Distribution of Y			
y_i	p_i	$p_i y_i$	$p_i y_i^2$
0	1/3	0	0
1	2/3	2/3	2/3
	3/3	2/3	2/3

$E[Y] = 2/3$

$$V[Y] = E[Y^2] - E[Y]^2$$
$$= 2/3 - (2/3)^2$$
$$= 2/9$$

Distribution of $Z = X - Y$			
z_i	p_i	$p_i z_i$	$p_i z_i^2$
0	2/12	0	0
1	7/12	7/12	7/12
2	3/12	6/12	12/12
	12/12	13/12	19/12

$E[Z] = 13/12$

$$V[Z] = E[Z^2] - E[Z]^2$$
$$= 19/12 - (13/12)^2$$
$$= 59/144$$

$$V[X] + V[Y] = 3/16 + 2/9$$
$$= 59/144$$

Therefore $\quad V[X] + V[Y] = V[X - Y]$

Exercise 3.2

1 The joint distribution of two discrete random variables X and Y is given by the following table.

		\multicolumn{3}{c}{x}
		1	2	3
	0	p	p	$2p$
y	1	p	$2p$	$3p$
	2	$2p$	$3p$	$5p$

a Find the value of p.
b Show that X and Y are not independent.
c Find $E[X]$ and $E[Y]$.
d Determine the distribution of $Z = XY$ and verify that $E[XY] \neq E[X].E[Y]$.

2 The discrete random variables X and Y have the joint distribution shown in the following table.

		\multicolumn{3}{c}{x}
		0	1	2
y	0	$2p$	$4p$	$2p$
	1	$3p$	$6p$	$3p$

a Find the value of p.
b Show that X and Y are independent.
c Find $E[X]$ and $E[Y]$.
d Determine the distribution of $Z = XY$ and verify that $E[XY] = E[X].E[Y]$.

3 The discrete random variables X and Y have the joint distribution shown in the following table.

		\multicolumn{3}{c}{x}
		0	1	2
	0	$2p$	p	$2p$
y	1	p	0	p
	2	$2p$	p	$2p$

a Find the value of p.
b Show that X and Y are not independent.
c Find $E[X]$ and $E[Y]$.
d Find the distribution of $Z = XY$ and verify that $E[XY] = E[X].E[Y]$.

3.3 Some properties of independent random variables

If X and Y are two independent discrete random variables then

$$E[XY] = E[X].E[Y] \tag{1}$$

$$V[aX + bY] = a^2V[X] + b^2V[Y] \tag{2}$$

The proofs of these results are given below.

$$E[XY] = \sum_{i=1}^{n} \sum_{j=1}^{m} p_{ij}x_iy_j$$

$$= \sum_{i=1}^{n} \sum_{j=1}^{m} p_ip'_jx_iy'_j$$

$$= \sum_{i=1}^{n} p_ix_i \left(\sum_{j=1}^{m} p'_jy_j \right)$$

$$= \sum_{i=1}^{n} p_ix_i \, E[Y]$$

$$= E[X].E[Y]$$

It should be noted that if $E[XY] = E[X].E[Y]$ it does not necessarily follow that X and Y are independent as was shown in Question **3** of Exercise 3.2.

Let $Z = aX + bY$, $\mu_X = E[X]$, $\mu_Y = E[Y]$ and $\mu_Z = E[Z]$.

$$E[Z] = E[aX + bY]$$

$$= aE[X] + bE(Y)$$

$$\therefore \quad \mu_Z = a\mu_X + b\mu_Y$$

$$Z - \mu_Z = (aX + bY) - (a\mu_X + b\mu_Y)$$

$$= a(X - \mu_X) + b(Y - \mu_Y)$$

$$V[Z] = E[(Z - \mu_Z)^2]$$

$$= E[a^2(X - \mu_X)^2 + 2ab(X - \mu_X)(Y - \mu_Y) + b^2(Y - \mu_Y)^2]$$

$$= a^2E[(X - \mu_X)^2] + 2abE[(X - \mu_X)(Y - \mu_Y)] + b^2E[(Y - \mu_Y)^2]$$

As X, Y independent, result (1) may be used on the middle term.

$$= a^2V[X] + 2abE[X - \mu_X].E[Y - \mu_Y] + b^2V[Y]$$

But $\quad E[X - \mu_X] = E[X] - \mu_X = 0$.

Therefore $\quad V[Z] = a^2V[X] + b^2V[Y]$.

Extensions to three or more discrete random variables

The following results concerning three or more discrete random variables may be proved from the above results.

$$E[a_1X_1 + a_2X_2 + \cdots + a_nX_n] = a_1E[X_1] + a_2E[X_2] + \cdots + a_nE[X_n]$$

When $X_1, X_2, \ldots, X_n$ are independent

$$V[a_1X_1 + a_2X_2 + \cdots + a_nX_n] = a_1^2V[X_1] + a_2^2V[X_2] + \cdots + a_n^2V[X_n]$$

and $\quad E[X_1X_2\ldots X_n] = E[X_1].E[X_2]\ldots E[X_n].$

Mean and variance of the binomial distribution

An alternative method of establishing the formulae for the mean and the variance of the binomial distribution uses the properties of the expected value and variance operators.

Let X be the number of successes obtained in a sequence of n independent Bernoulli trials in each of which the probability of a success is p.

$$X \sim B(n, p)$$

Let X_i be the number of successes obtained in the ith trial. The distribution of X_i is given by the first two columns of the following table. The values of $E[X_i]$ and $V[X_i]$ are found in the usual way.

x_i	p_i	p_ix_i	$p_ix_i^2$
0	$1 - p$	0	0
1	p	p	p
	1	p	p

$$E[X_i] = p$$
$$V[X_i] = E[X_i^2] - (E[X_i])^2$$
$$= p - p^2$$
$$= p(1 - p)$$

These results are true for $i = 1, 2, \ldots, n$.

The total number of successes in the n trials is obtained by adding the number of successes in each of the trials.

$$X = X_1 + X_2 + \cdots + X_n$$
$$E[X] = E[X_1 + X_2 + \cdots + X_n]$$
$$= E[X_1] + E[X_2] + \cdots + E[X_n]$$
$$= p + p + \cdots + p \qquad (n \text{ times})$$

Therefore $\quad E[X] = np.$

$$V[X] = V[X_1 + X_2 + \cdots + X_n]$$

Since $X_1, X_2, \ldots, X_n$ are independent

$$V[X] = V[X_1] + V[X_2] + \cdots + V[X_n]$$
$$= p(1 - p) + p(1 - p) + \cdots + p(1 - p) \qquad (n \text{ times})$$

Therefore $\quad V[X] = np(1 - p).$

Thus the mean and the variance of $B(n, p)$ are np and $np(1 - p)$ respectively.

43

Additive property of independent Poisson distributions

If X and Y are independent random variables having Poisson distributions with means α and β respectively, then $Z = X + Y$ has a Poisson distribution with mean $\alpha + \beta$.

Proof

$$P(Z = r) = \sum_{s=0}^{r} P(X = s \cap Y = r - s)$$

$$= \sum_{s=0}^{r} P(X = s) \cdot P(Y = r - s) \qquad (X, Y \text{ independent})$$

$$= \sum_{s=0}^{r} e^{-\alpha} \frac{\alpha^s}{s!} \cdot e^{-\beta} \frac{\beta^{(r-s)}}{(r - s)!}$$

$$= e^{-(\alpha + \beta)} \sum_{s=0}^{r} \frac{\alpha^s}{s!} \cdot \frac{\beta^{(r-s)}}{(r - s)!}$$

$$= \frac{e^{-(\alpha + \beta)}}{r!} \sum_{s=0}^{r} \frac{r! \alpha^s \beta^{(r-s)}}{s!(r - s)!}$$

$$= \frac{e^{-(\alpha + \beta)}}{r!} \sum_{s=0}^{r} \binom{r}{s} \alpha^s \beta^{(r-s)}$$

$$= \frac{e^{-(\alpha + \beta)}}{r!} (\alpha + \beta)^r \qquad (\text{binomial theorem})$$

This is true for $r = 0, 1, 2, \ldots$.

Therefore $\qquad Z \sim \text{Po}(\alpha + \beta)$.

This result may be extended to include three or more independent Poisson variables.

Example 6

X and Y are independent discrete random variables such that
$X \sim B(100, 0.64)$ and $Y \sim \text{Po}(20)$. Find the mean and variance of $X - 2Y$

Since $X \sim B(100, 0.64)$ $\qquad E[X] = 100 \times 0.64 = 64$

$\qquad\qquad\qquad\qquad\qquad V[X] = 100 \times 0.64 \times (1 - 0.64) = 23.04$

Since $Y \sim \text{Po}(20)$ $\qquad\qquad E[Y] = 20$

$\qquad\qquad\qquad\qquad\qquad V[Y] = 20$

$$E[X - 2Y] = E[X] - 2E[Y]$$
$$= 64 - 2 \times 20$$
$$= 24$$

$$V[X - 2Y] = V[X] + 4V[Y]$$
$$= 23.04 + 4 \times 20$$
$$= 103.04$$

Example 7

A secretary is typing a document which has ten pages. Independently for each page, the number of mistakes made has a Poisson distribution with mean 0.25. Find the probability that the total number of mistakes made is less than six.

Let $X_1, X_2, \ldots, X_{10}$ be the number of mistakes made on the 1st, 2nd, ..., 10th pages respectively.

$$X_1 \sim \text{Po}(0.25), \; X_2 \sim \text{Po}(0.25), \; \ldots, \; X_{10} \sim \text{Po}(0.25)$$

Let X be the total number of mistakes made.

$$X = X_1 + X_2 + \ldots + X_{10}$$

Since $X_1, X_2, \ldots, X_{10}$ are independent, X is a Poisson variable with mean equal to 10×0.25.

$$X \sim \text{Po}(2.5)$$

$$\text{P}(X < 6) = \text{P}(X \leqslant 5)$$
$$= 0.958 \quad \text{(from \textbf{Table 2}, page 230)}$$

Exercise 3.3

1   Given that $X \sim \text{B}(20, 0.6)$, $Y \sim \text{B}(10, 0.4)$ and that X and Y are independent, find the mean and variance of

   a $X + Y$ **b** $X - Y$ **c** $3X + 2Y$.

2   Given that $X \sim \text{Po}(6)$, $Y \sim \text{Po}(4)$ and that X and Y are independent, find the mean and variance of

   a $X + Y$ **b** $X - Y$ **c** $3X + 2Y$.

3   Telephone calls reach a secretary independently and at random, internal calls arrive at a mean rate of two in any five-minute period and external calls arrive at a mean rate of one in any five-minute period. Find the probability that there will be more than four calls received by the secretary in any five-minute period.

4   In a cafeteria the numbers of cups of coffee and tea sold per minute may be assumed to be independent Poisson variables with means 2 and 1 respectively.

   a Calculate the probability that in a given one-minute period exactly two cups of coffee and one cup of tea are sold.

   b Find the probability that in a given three-minute period fewer than ten drinks altogether are sold.

5   The numbers, X and Y, of emissions per minute from two independent radioactive sources have Poisson distributions with means 3 and 7 respectively.

   a Find the probability that in any minute the total number of emissions from the two sources is more than four.

   b Find the mean and variance of $3X - Y$.

Miscellaneous Exercise 3

1 The two discrete random variables X and Y have the joint probability distribution displayed in the following table.

			x	
		0	1	2
y	0	1/5	1/10	11/30
	1	1/20	3/20	2/15

(i) Determine whether or not X and Y are independent.

(ii) Show that $E[XY] = E[X].E[Y]$. *(WJEC)*

2 A factory makes fruit gums in the three flavours 'orange', 'lemon' and 'strawberry'. The proportions of gums that are orange-flavoured, lemon-flavoured and strawberry-flavoured are 0.4, 0.2 and 0.4, respectively. In a random sample of three of these fruit gums, let X denote the number that are orange-flavoured and let Y denote the number that are lemon-flavoured.

(i) Name the distributions of X and Y, and write down their means and variances.

(ii) The following table displays the joint probability distribution of X and Y. Verify that the entry $P(X = 1, Y = 2) = 0.048$ is correct, and calculate the five missing entries in the table.

				x	
		0	1	2	3
y	0	0.064	0.192		
	1	0.096		0.096	0
	2		0.048	0	0
	3		0	0	0

(iii) Calculate the probability that a random sample of three gums will include more strawberry-flavoured gums than the combined number of orange-flavoured and lemon-flavoured gums. *(WJEC)*

3 The two discrete random variables X and Y have the joint probability distribution displayed in the following table.

			x	
		0	1	2
y	1	0	0.05	0
	2	0.2	0	0.15
	3	0	0.6	0

(i) Show that X and Y are not independent.

(ii) Evaluate $E[XY]$.

(iii) Derive the joint probability distribution of $U = Y + X$ and $W = Y - X$. Show that U and W are independent. Express $Y - 3X$ in terms of U and W and hence evaluate $V[Y - 3X]$.   *(WJEC)*

4 Four balls are to be drawn at random, without replacement, from a bag containing 10 balls, of which 2 are red, 1 is blue and 7 are white. Show that the probability that all four balls drawn will be white is equal to $\dfrac{1}{6}$.

Let X and Y, respectively, denote the numbers of red and blue balls that will be drawn. Display the joint probability distribution of X and Y in a two-way table. Use entries from your table to show that X and Y are not independent.

Find the probability distribution of $Z = X + Y$, and evaluate its mean and variance.

If, instead of as above, the four balls were to be drawn at random with replacement, identify the distribution of $Z = X + Y$ by name and write down the values of its mean and variance.   *(WJEC)*

5 A boy is to have three attempts at performing a certain task. The probability that he will succeed on his first attempt is $\dfrac{1}{2}$, while on each subsequent attempt the probability that he will succeed is $\dfrac{3}{4}$ if his preceding attempt was successful, and $\dfrac{1}{2}$ if his preceding attempt was not successful. Let X denote the number of successes by the boy in his three attempts, and let Y denote the number of the attempt on which he is first successful; define Y to be zero if the boy fails on all three attempts.

(i) Display the joint probability distribution of X and Y in a two-way table.

(ii) Having completed his three attempts, the boy is awarded a score of 2 if $X > Y$, a score of 1 if $X = Y$ and a score of 0 if $X < Y$. Find the mean and the standard deviation of the boy's score.   *(WJEC)*

6 Given that X and Y are independent and have Poisson distributions with means a and b respectively, show that the sum of these quantities $X + Y$, has probabilities of values 0, 1, 2 which are in accordance with those of a Poisson distribution with mean $a + b$.

A shopkeeper has two shops which are supplied from a central store. For a particular product each shop asks the store for a complete box when required. The numbers of boxes per week requested by the two shops are independent and have Poisson distributions with means $\frac{4}{5}$ and $\frac{1}{5}$ respectively. Find the probabilities that two or more boxes are requested from the store in a week

(i) by the first shop,

(ii) by the second shop,

(iii) altogether.

Calculate the lowest level of stock at the central store for which there is a probability greater than 90% that all demands from the shops in the next week can be met. (*JMB*)

7 Derive the mean and the variance of a Poisson distribution.

Two types of flaws, *A* and *B*, may occur in a manufactured cloth. The numbers of flaws of type *A* and of type *B* occurring per metre length of the cloth are independent random variables having Poisson distributions with means 0.5 and 1, respectively.

a Find the probabilities, to three significant figures, that a length of 1 metre of the cloth will have

(i) 2 or fewer flaws of type *A*,

(ii) no flaw of either type.

b Show that the probability of a length of 1 metre of the cloth containing 1 flaw only is exactly three times that of it containing 1 flaw of each type.

c Removing a type *A* flaw from the cloth costs 8 pence and removing a type *B* flaw costs 2 pence. Find the mean and the standard deviation of the cost of removing flaws per 1 metre length of cloth. (*WJEC*)

8 A college's switchboard handles both internal and external telephone calls. In any half-hour period the numbers of internal and external calls are independent and have Poisson distributions with means λ and μ, respectively.

a Find expressions, in terms of λ and μ, for the probabilities that during a half-hour period there will be

(i) no call at all,

(ii) 2 internal calls and 1 external call,

(iii) a total of 3 calls.

b Given that $\lambda = 3$ and $\mu = 5$, obtain the probabilities that during a half-hour period there will be

(i) at least 10 internal calls,

(ii) exactly 3 external calls.

(iii) Given that a total of 3 calls arrived during a half-hour period find the conditional probability that exactly 2 of them were internal calls. (*WJEC*)

9 A random variable X has a Poisson distribution with parameter μ, so that

$$P(X = r) = e^{-\mu}\frac{\mu^r}{r!} \qquad (r = 0, 1, 2, \ldots).$$

Derive the mean and the variance of X.

The weekly sales X of a particular type of item A by a shop has a Poisson distribution with parameter 9.8. Find the probability that more than 10 items will be sold in a week.

The weekly sales Y of another type of item B by the same shop has a Poisson distribution with parameter 3.8. Assuming that X and Y are independent find the probability that the shop will sell exactly 5 items of type A and exactly 5 items of type B in a week.

Suppose the shop makes a profit of 10 pence on every A item sold and a profit of 20 pence on every B item sold. Calculate the mean and the standard deviation of the total weekly profit made by the shop from the sale of items of type A and type B. *(WJEC)*

10 Let X denote the number of heads obtained in two tosses of a fair coin, and let Y denote a random variable which is independent of X and has a Poisson distribution with mean 1.

  (i) Find, in terms of e, the probability that both X and Y will have the value zero.

  (ii) Find, in terms of e, the probability that X and Y will have the same value.

Given that $Z = XY$,

  (iii) express $P(Z = 0)$ in terms of e,

  (iv) evaluate the mean and the variance of Z. *(WJEC)*

Chapter 4

Joint distributions of continuous random variables

The result
$$E[a_1X_1 + a_2X_2 + \cdots + a_nX_n] = a_1E[X_1] + a_2E[X_2] + \cdots + a_nE[X_n]$$
and the results, when $X_1, X_2, \ldots, X_n$ are independent,
$$V[a_1X_1 + a_2X_2 + \cdots + a_nX_n] = a_1{}^2V[X_1] + a_2{}^2V[X_2] + \cdots + a_n{}^2V[X_n]$$
$$E[X_1X_2\ldots X_n] = E[X_1]E[X_2]\ldots E[X_n].$$
are also applicable when some, or all, of $X_1, X_2, \ldots, X_n$ are continuous random variables. The proofs of these results involve mathematics beyond A level.

4.1 Linear combinations of independent normal variables

It may be shown that any linear combination of independent normal variables is itself normally distributed.

If $\quad X_1 \sim N(\mu_1, \sigma_1{}^2), X_2 \sim N(\mu_2, \sigma_2{}^2), \ldots, X_n \sim N(\mu_n, \sigma_n{}^2)$ are independent

and $\quad Y = a_1X_1 + a_2X_2 + \cdots + a_nX_n$

then $\quad Y \sim N(\mu, \sigma^2)$

where $\quad \mu = a_1\mu_1 + a_2\mu_2 + \cdots + a_n\mu_n$

and $\quad \sigma^2 = a_1{}^2\sigma_1{}^2 + a_2{}^2\sigma_2{}^2 + \cdots + a_n{}^2\sigma_n{}^2.$

Example 1

The weights of male and female students at a college are normally distributed, the male weights having a mean of 70 kg and a standard deviation of 8 kg and the female weights having a mean of 50 kg and a standard deviation of 6 kg. If one male student and one female student are chosen at random, calculate the probabilities that

a   the sum of their weights is less than 130 kg

b   the female student is heavier than the male student

c   the female student's weight is at least 90% of the male student's weight.

a If M denotes the weight of the male student, then $M \sim N(70, 8^2)$.

If F denotes the weight of the female student, then $F \sim N(50, 6^2)$.

Let $S = M + F$

$$E[S] = E[M] + E[F] \quad \text{and} \quad V[S] = V[M] + V[F]$$
$$E[S] = \quad 70 + \quad 50 \qquad\qquad V[S] = \quad 8^2 + 6^2$$
$$E[S] = 120 \qquad\qquad\qquad\qquad V[S] = 100$$
$$S \sim N(120, 100)$$

Let $Z = \dfrac{S - 120}{10}$ then $Z \sim N(0, 1)$

$$P(S < 130) = P\left(Z < \frac{130 - 120}{10}\right)$$
$$= P(Z < 1)$$
$$= 0.8413 \qquad\qquad \text{(from \textbf{Table 3}, page 231)}$$

b It is required to find $P(F > M)$, i.e. $P(F - M > 0)$. Consider the random variable $D = F - M$.

$$E[D] = E[F] - E[M] \quad \text{and} \quad V[D] = V[F] + V[M]$$
$$E[D] = \quad 50 - \quad 70 \qquad\qquad V[D] = \quad 6^2 + 8^2$$
$$E[D] = -20 \qquad\qquad\qquad\qquad V[D] = 100$$
$$D \sim N(-20, 100)$$

Let $Z = \dfrac{D - (-20)}{10}$ then $Z \sim N(0, 1)$

$$P(D > 0) = P\left(Z > \frac{0 + 20}{10}\right)$$
$$= P(Z > 2)$$
$$= 0.0228 \qquad\qquad \text{(from \textbf{Table 3})}$$

c It is required to find $P(F > 0.9M)$, i.e. $P(F - 0.9M > 0)$. Consider the random variable $Y = F - 0.9M$.

$$E[Y] = E[F] - 0.9E[M] \quad \text{and} \quad V[Y] = V[F] + 0.9^2 V[M]$$
$$E[Y] = \quad 50 - 0.9 \times 70 \qquad\qquad V[Y] = \quad 6^2 + 0.9^2 \times 8^2$$
$$E[Y] = -13 \qquad\qquad\qquad\qquad V[Y] = 87.84$$
$$Y \sim N(-13, 87.84)$$

Let $Z = \dfrac{Y - (-13)}{\sqrt{87.84}}$ then $Z \sim N(0, 1)$

$$P(Y > 0) = P\left(Z > \frac{0 + 13}{\sqrt{87.84}}\right)$$
$$= P(Z > 1.387)$$
$$= 0.083 \quad \text{(3 d.p.)} \qquad\qquad \text{(from \textbf{Table 3})}$$

STATISTICS 2

Example 2

The weights of oranges are normally distributed with mean 195 g and standard deviation 10 g. A supermarket sells the oranges in packs each containing four randomly selected oranges. Find the probability that the weight of the oranges in one of these packs exceeds 750 g.

Let X_1, X_2, X_3, X_4 be the weights of the four oranges. X_1, X_2, X_3, X_4 are four independent random variables, each normally distributed with mean 195 g and standard deviation 10 g.

Let $Y = X_1 + X_2 + X_3 + X_4$

$E[Y]=E[X_1]+E[X_2]+E[X_3]+E[X_4]$ and $V[Y]=V[X_1]+V[X_2]+V[X_3]+V[X_4]$

$E[Y]= 195 + 195 + 195 + 195$ $\qquad$ $V[Y]= 10^2 + 10^2 + 10^2 + 10^2$

$E[Y]= 780$ $\qquad$ $V[Y]= 400$

$$Y \sim N(780, 400)$$

Let $Z = \dfrac{Y - 780}{20}$ then $Z \sim N(0, 1)$

$$P(Y > 750) = P\left(Z > \frac{750 - 780}{20}\right)$$
$$= P(Z > -1.5)$$
$$= 0.9332 \qquad \text{(from \textbf{Table 3})}$$

Exercise 4.1

1 The heights of men in a certain town are normally distributed with mean 172 cm and standard deviation 6 cm; the heights of women in the town are normally distributed with mean 160 cm and standard deviation 4 cm. A man and a woman are chosen at random. Find the probabilities that

a the woman is taller than the man,

b the woman's height is greater than 95% of the man's height.

2 The weights of the luggage of passengers using a certain airline are normally distributed with mean 18 kg and standard deviation 2 kg. Find the probability that the total weight of the luggage of 196 passengers on a particular flight will be between 3500 kg and 3600 kg.

3 The weight F of a full tin of rice pudding produced by a certain manufacturer is normally distributed with mean 560 g and standard deviation 13 g. The weight E of an empty tin used by the manufacturer is normally distributed with mean 30 g and standard deviation 5 g.

a Find the mean and standard deviation of the weight C of the contents of the tin, stating any assumption made.

b Determine the probability that the weight of the contents of a randomly chosen tin will be less than 500 g.

4 The times taken by two runners A and B to run 400 m races are independent and normally distributed with means 51.8 s and 52.3 s, and standard deviations 0.3 s and 0.4 s respectively. The two runners compete in a 400 m race. Find the probability that A will beat B.

5 A doctor finds that the consulting times of his patients for morning surgery are independent and normally distributed with mean 10 minutes and standard deviation 2 minutes. He sees his patients consecutively with no gap between consultations, starting at 9.00 a.m. At what time (to the nearest minute) should the ninth patient arrange for a taxi to call at the surgery so as to be 99% certain that the taxi is not kept waiting.

Miscellaneous Exercise 4

1 The times taken by two runners A and B to run 400 m races are independent and normally distributed with means 45.0 s and 45.2 s, and standard deviations 0.5 s and 0.8 s respectively. The two runners are to compete in a 400 m race for which there is a track record of 44.5 s.
  (i) Calculate, to three decimal places, the probability of runner A breaking the track record.
  (ii) Show that the probability of runner B breaking the track record is greater than that of runner A.
  (iii) Calculate, to three decimal places, the probability of runner A beating runner B. (*JMB*)

2 In a certain country, the heights of men are normally distributed with mean 172 cm and standard deviation 4 cm; and the heights of women are normally distributed with mean 164 cm and standard deviation 3 cm. From this country one man and one woman are chosen randomly. Assuming that a linear combination of independent normally distributed random variables is itself normally distributed, calculate, correct to three decimal places, the probabilities that
  (i) the man is taller than the woman,
  (ii) the woman's height is at least 90 per cent of the man's height. (*JMB*)

3 The internal diameters of circular tubes from a certain manufacturer are distributed normally with mean 30 mm. It is observed that 97.5 per cent of the tubes have internal diameters greater than 29.02 mm. Show that the proportion of tubes with internal diameters greater than 31 mm is 0.0228, correct to four decimal places.

The manufacturer also produces circular rods, the diameters of which are normally distributed with standard deviation 1.50 mm. Given that 97.5 per cent have diameters less than 29.94 mm, find the proportion that have diameters less than 28.50 mm.

If a randomly chosen tube has internal diameter X mm and a randomly chosen rod has diameter Y mm, state the probability distribution of $X - Y$. Hence find the probability that a randomly chosen rod will fit inside a randomly chosen tube. (*JMB*)

STATISTICS 2

4 The masses of packets of sweets of a particular brand are distributed normally with mean 245 g and standard deviation 2 g. Calculate, to three decimal places, the probabilities that the mean mass of four randomly chosen packets
(i) will exceed 247 g,
(ii) will be within 1 g of 245 g.

Let X_1, X_2, X_3, respectively, denote the masses in grams of three randomly chosen packets. Find the mean and the variance of $X_1 + X_2 - 2X_3$. *(JMB)*

5 The weights of the contents of cans of fruit are normally distributed with mean 250.2 g and standard deviation 2 g.
(i) Calculate the proportion of the cans that contain less than 250 g.
(ii) Given that 75% of the cans contain at least w g, find the value of w correct to one decimal place.
(iii) Find the probability that the combined contents of four cans will weigh more than 1 kg.
(iv) The weights of the filled cans are distributed with mean 274.5 g and standard deviation 2.5 g. Assuming that the weight of the contents of a can is independent of the weight of the can when empty, determine the mean and the standard deviation of the weights of the empty cans. *(WJEC)*

6 The tensile strengths, measured in newtons (N), of a large number of ropes of equal length are independently and normally distributed such that five per cent are under 706 N and five per cent over 1294 N. Four such ropes are randomly selected and joined end-to-end to form a single rope; the strength of the combined rope is equal to the strength of the weakest of the four selected ropes. Derive the probabilities that this combined rope will not break under tensions of 1000 N and 900 N, respectively.

A further four ropes are randomly selected and attached between two rings, the strength of the arrangement being the sum of the strengths of the four separate ropes. Derive the probabilities that this arrangement will break under tensions of 4000 N and 4200 N, respectively.

Find the smallest number of ropes that should be selected if the probability that at least one of them has a strength greater than 1000 N is to exceed 0.99. *(JMB)*

7 Weights of persons using a certain lift are normally distributed with mean 150 lb and standard deviation 20 lb. The lift has a maximum permissible load of 650 lb.
(i) Four persons arising at random from this population are in the lift. Determine the probability that the maximum load is exceeded.
(ii) One person arising at random from this population is in the lift. He is accompanied by luggage weighing three times his own weight. Determine the probability that the maximum load is exceeded.

Explain any difference in your answers to (i) and (ii). *(JMB)*

8 In testing the length of life of electric light bulbs of a particular type, it is found that 12.3% of the bulbs tested fail within 800 hours and that 28.1% are still operating 1100 hours after the start of the test. Assuming that the distribution of the length of life is normal, calculate, to the nearest hour in each case, the mean, μ, and the standard deviation, σ, of the distribution.

A light-fitting takes a single bulb of this type. A packet of three bulbs is bought, to be used one after the other in the fitting. State the mean and variance of the total life of the three bulbs in the packet in terms of μ and σ and calculate, to two decimal places, the probability that the total life is more than 3300 hours.

Calculate the probability that all three bulbs have lives in excess of 1100 hours, so that again the total life is more than 3300 hours. Explain why this answer should be different from the previous one. *(JMB)*

9 A long-distance lorry driver works a nine-hour day, including time for a lunch break. On each day, he sets off and drives for 4 hours; he then stops for a lunch break of 1 hour, after which he continues the journey for a further 4 hours. Independently for each day, the average speeds of the lorry before and after lunch are independent random variables each having a normal distribution with mean 60 km/h and standard deviation 10 km/h.
  (i)   Show that the probability that the driver will have travelled at least 200 km when he stops for his lunch break is equal to 0.841, correct to three decimal places.
  (ii) Find the value of s, correct to three significant figures, if there is a probability of 0.9 that the driver will have travelled at least s km when he stops for his lunch break.
  (iii) Find, correct to three decimal places, the probability that the distance travelled after the lunch break will be at least 10 km more than that travelled before the lunch break.
  (iv) Find, correct to three decimal places, the probability that the total distance travelled over two successive days will be less than 800 km.
                                                                      (WJEC)

10 A box contains three electric light bulbs of brand A and two of brand B. Bulbs of brand A have lifetimes that are normally distributed with mean 1200 hours and standard deviation 200 hours, while bulbs of brand B have lifetimes that are normally distributed with mean 1400 hours and standard deviation 400 hours. Two bulbs are selected at random without replacement from the box. Calculate, to three significant figures,
  (i)   the probability that the first bulb selected will have a lifetime in excess of 1300 hours,
  (ii) the probability that the sum of the lifetimes of the two selected will exceed 3000 hours. *(JMB)*

Chapter 5

Sampling distributions

In a statistical experiment the universal set of the objects under investigation is called a *population*. The probability distribution of a random variable X associated with the objects in the population is called the *population distribution* of X; this distribution has certain parameters which may be unknown. For example, in a quality control experiment when the population is the entire output of a factory producing TV tubes, X might be the operational lifetime of a randomly chosen TV tube and the mean μ of the operational lifetimes of all the TV tubes produced one of the unknown parameters of the population distribution of X.

In many experiments it is not possible to obtain information about every member of the population; in the example above, determining the lifetime of every TV tube would involve the destruction of the entire output of the factory; in other examples, considerations of time and expense may prohibit examining the entire population. In such cases it is usual to take a *sample*, which is a subset of the population, and from calculations based on the observations of the members of the sample to draw conclusions about the population as a whole. Any quantity calculated from the sample values for the purpose of estimating a population parameter is called a *statistic*; in the above example, the mean operational lifetime of a sample of 20 TV tubes could be used as a statistic to estimate μ. A statistic is a random variable having its own distribution which is called the *sampling distribution* of that statistic.

In order that inferences drawn from statistical experiments of this nature should be reliable, it is important that the samples studied should be representative of the population; this is often achieved by selecting a *simple random sample*, defined below.

A random sample of size n from a finite population is a selection of n members of the population made in such a way that all possible selections of n members are equally likely to be chosen.

This definition covers sampling both with and without replacement.

5.1 Sampling without replacement from a finite population

When the number of objects in the population and the sample size are both small, it is feasible to list all possible samples and by calculating the value of a statistic for each sample the sampling distribution of the statistic may be obtained. The following example illustrates this method in the case when the sampling is performed without replacement.

Example 1

Schools A, B, C, D, E have 4, 4, 4, 6, 8 full-time Mathematics teachers respectively. Let X denote the number of full-time Mathematics teachers in a randomly chosen school. Find the mean μ of the population distribution of X.

A sample of two schools is chosen at random without replacement. Let $\bar{X}$ denote the mean number of full-time Mathematics teachers per school in the sample. Find the sampling distribution of $\bar{X}$ and calculate its mean and standard deviation.

The population distribution of X is given by the first two columns in the following table:

x_i	p_i	$p_i x_i$
4	3/5	12/5
6	1/5	6/5
8	1/5	8/5
	5/5	26/5

the population mean $\mu = 26/5 = 5.2$.

The following table lists all possible equally likely samples of size 2 from the 5 schools, together with the sample means of their X-values.

Sample		X-values		$\bar{X}$
A	B	4	4	4
A	C	4	4	4
A	D	4	6	5
A	E	4	8	6
B	C	4	4	4
B	D	4	6	5
B	E	4	8	6
C	D	4	6	5
C	E	4	8	6
D	E	6	8	7

The sampling distribution of $\bar{X}$ is given in the first two columns of the table below:

$\bar{x}_i$	p_i	$p_i \bar{x}_i$	$p_i \bar{x}_i^2$
4	3/10	12/10	48/10
5	3/10	15/10	75/10
6	3/10	18/10	108/10
7	1/10	7/10	49/10
	10/10	52/10	280/10

$$E[\bar{X}] = 52/10 = 5.2$$
$$V[\bar{X}] = E[\bar{X}^2] - \{E[\bar{X}]\}^2$$
$$= 28 - 5.2^2$$
$$= 0.96$$
$$SD[\bar{X}] = \sqrt{0.96} = 0.980 \quad (3 \text{ d.p.})$$

It is intuitively plausible to use the statistic $\bar{X}$ as an estimator for the population mean μ. Since, in this example, the population distribution and the sampling distribution of $\bar{X}$ are both known, certain aspects of the use of $\bar{X}$ as an estimator of μ may be examined. Although $\bar{X}$ (which takes the values 4, 5, 6, 7) will never equal μ (which is 5.2), it should be noted that $E[\bar{X}] = \mu$. This property of an estimator, i.e. that 'on average' the estimator gives the true value of the quantity being estimated, is a desirable one and an estimator having this property is said to be *unbiased*. The result, $E[\bar{X}] = \mu$, is true in the more general case when $\bar{X}$ is the sample mean of random samples of any size drawn from a population. Thus, in a more realistic version of **Example 1**, when a random sample of 20 schools is chosen from all the secondary schools in a certain area the sample mean number of full-time Mathematics teachers per school is an unbiased estimate of the population mean.

Since $E[\bar{X}] = \mu$, an appropriate measure of the closeness of $\bar{X}$ to μ is the standard deviation of the sampling distribution of $\bar{X}$. The standard deviation of a sampling distribution of a statistic is often referred to as the *standard error* of the statistic, as the smaller is its value the more likely is the value of the statistic to be near the quantity being estimated. If **Example 1** were repeated using samples of size 3 instead of size 2, the values for $E[\bar{X}]$ and $SD[\bar{X}]$ that would be obtained are 5.2 and 0.653 respectively (see Exercise 5.1, question **1**). It should be noted that $\bar{X}$ is an unbiased estimator of μ and that the standard error of $\bar{X}$ is smaller when the sample size is larger.

Exercise 5.1

1 Schools A, B, C, D, E have 4, 4, 4, 6, 8 full-time Mathematics teachers respectively. A sample of the three schools is chosen at random without replacement from the five schools. Let $\bar{X}$ denote the mean number of full-time Mathematics teachers per school in the sample. Find the sampling distribution of $\bar{X}$, and calculate its mean and standard deviation.

2 A pack consists of five cards numbered 1, 2, 3, 4, 5 respectively. Let X denote the number of a card chosen at random from the pack. Find the mean μ of the probability distribution of X.

Three cards are chosen at random without replacement from the pack. Let M denote the median of the numbers on the chosen cards. Find the sampling distribution of M and determine whether M is an unbiased estimator of μ.

3 A population consists of the six numbers 1, 2, 2, 2, 3, 3. Two numbers are selected at random without replacement. Find the sampling distribution of the larger of the two numbers selected.

4 A population consists of the five numbers 0, 1, 2, 3, 4. Find the population mean μ and variance σ^2.

Two numbers are chosen at random without replacement from the population. Let V denote the variance of the pair of chosen numbers. Find the sampling distribution of V and show that $E[V] \neq \sigma^2$.

5 A bag contains 3 red balls and 2 blue balls. Two balls are chosen at random without replacement from the bag. Let Y denote the proportion of red balls in the sample. Find the sampling distribution of Y and verify that Y is an unbiased estimator for the proportion of red balls in the bag.

5.2 Sampling from a distribution

Before discussing sampling from a distribution in general, a special case will be considered.

Sampling with replacement from a finite population

Example 2

Schools A, B, C, D, E have 4, 4, 4, 6, 8 full-time Mathematics teachers. A sample of two schools is chosen at random with replacement from the five schools. Let $\bar{X}$ denote the mean number of full-time Mathematics teachers per school in the sample. Find the sampling distribution of $\bar{X}$, and calculate its mean and variance.

Instead of writing down the 25 equally likely samples of two schools and proceeding as in **Example 1**, an alternative method will be adopted. In the following table all possible samples of the X-values are recorded together with their associated probabilities (since they are not all equally likely).

Sample	Probability		$\bar{X}$
4 4	$\frac{3}{5} \times \frac{3}{5}$	$= \frac{9}{25}$	4
4 6	$\frac{3}{5} \times \frac{1}{5} \times 2$	$= \frac{6}{25}$	5
4 8	$\frac{3}{5} \times \frac{1}{5} \times 2$	$= \frac{6}{25}$	6
6 6	$\frac{1}{5} \times \frac{1}{5}$	$= \frac{1}{25}$	6
6 8	$\frac{1}{5} \times \frac{1}{5} \times 2$	$= \frac{2}{25}$	7
8 8	$\frac{1}{5} \times \frac{1}{5}$	$= \frac{1}{25}$	8

STATISTICS 2

The sampling distribution of $\bar{X}$ is given by the first two columns of the following table.

$\bar{x}_i$	p_i	$p_i\bar{x}_i$	$p_i\bar{x}_i^2$
4	9/25	36/25	144/25
5	6/25	30/25	150/25
6	7/25	42/25	252/25
7	2/25	14/25	98/25
8	1/25	8/25	64/25
	25/25	130/25	708/25

$$E[\bar{X}] = 130/25 = 5.2$$
$$V[\bar{X}] = E[\bar{X}^2] - \{E[\bar{X}]\}^2$$
$$= 708/25 - 5.2^2$$
$$= 1.28$$

The values of the population mean and variance, which were calculated in **Example 1**, are given by $\mu = 5.2$ and $\sigma^2 = 2.56$.

It should be noted that $E[\bar{X}] = \mu$, so that $\bar{X}$ is an unbiased estimator of μ and that $V[\bar{X}] = \dfrac{\sigma^2}{n}$, where n is the sample size, (2 in this case).

It should also be noted that $V[\bar{X}]$, 0.96, in **Example 1** (when the sampling was done without replacement) is smaller than $V[\bar{X}]$, 1.28, in **Example 2** (when sampling was done with replacement), indicating that the former method is to be preferred.

Random sample from a distribution

The terms 'sample' and 'population' may be used in circumstances where no collection of objects actually exists. For example, when a fair die is thrown continually an infinite sequence of scores, each of which will be one of the numbers 1, 2, 3, 4, 5, 6, will be obtained. This infinite sequence of numbers may be defined to be a population. The score on one throw has a discrete uniform distribution, which may be referred to as the population distribution. The scores obtained in a finite number n of throws may be regarded as a sample of size n from the population distribution. The formal definition of a *random sample from a distribution* is given below.

A random sample of size n from the population distribution of X is a set of independent random variables $X_1, X_2, \ldots, X_n$, each of which has the same distribution as X.

The set $\{X_1, X_2, \ldots, X_n\}$ is also called *a random sample of n observations of X* or *a sample of n independent observations of X*.

Random sampling with replacement from a finite population is a special case, since, on each occasion a member of the sample is chosen, the selection is performed under exactly the same circumstances. On the other hand, the conditions stated above are not satisfied when sampling is performed without replacement, as the probability of any specific selection is affected by the outcomes of previous selections.

Distribution of a sample mean

Let X be a random variable with a distribution having mean μ and variance σ^2. Let $X_1, X_2, \ldots, X_n$ be a random sample of size n from the distribution of X. Let $\bar{X}$ be the sample mean.

$$E[\bar{X}] = E\left[\frac{X_1 + X_2 + \cdots + X_n}{n}\right]$$

$$= \frac{1}{n}E[X_1 + X_2 + \cdots + X_n]$$

$$= \frac{1}{n}(E[X_1] + E[X_2] + \cdots + E[X_n])$$

Since each X_i has the same distribution as X, $E[X_i] = \mu$ $(i = 1, 2, \ldots, n)$.

$$E[\bar{X}] = \frac{1}{n}(\mu + \mu + \cdots + \mu)$$

$$= \frac{1}{n} \times n\mu$$

$$E[\bar{X}] = \mu$$

$$V[\bar{X}] = V\left[\frac{X_1 + X_2 + \cdots + X_n}{n}\right]$$

$$= \frac{1}{n^2}V[X_1 + X_2 + \cdots + X_n]$$

Since $X_1, X_2, \ldots, X_n$ are independent.

$$V[\bar{X}] = \frac{1}{n^2}(V[X_1] + V[X_2] + \cdots V[X_n])$$

Since each X_i has the same distribution as X, $V[X_i] = \sigma^2$ $(i = 1, 2, \ldots, n)$.

$$V[\bar{X}] = \frac{1}{n^2}(\sigma^2 + \sigma^2 + \cdots + \sigma^2)$$

$$= \frac{1}{n^2} \times n\sigma^2$$

$$V[\bar{X}] = \frac{\sigma^2}{n}$$

The results $E[\bar{X}] = \mu$ and $V[\bar{X}] = \frac{\sigma^2}{n}$ have a very wide application as they are valid for any distribution of X, provided that each member of the sample was independently selected under identical circumstances.

Example 3

State the values of the mean and the variance of the continuous random variable X which is uniformly distributed between 2 and 8. Find the mean and the standard error of the mean $\bar{X}$ of a random sample of 48 observations of X.

$$X \sim U(2, 8)$$

$$\mu = \frac{2 + 8}{2} \qquad \sigma^2 = \frac{(8 - 2)^2}{12}$$

$$\mu = 5 \qquad \sigma^2 = 3$$

$$E[\bar{X}] = \mu \qquad V[\bar{X}] = \frac{\sigma^2}{n} = \frac{3}{48} = \frac{1}{16}$$

$$E[\bar{X}] = 5 \qquad SE[\bar{X}] = \frac{1}{4}$$

Exercise 5.2

1 Schools A, B, C, D, E have 4, 4, 4, 6, 8 full-time Mathematics teachers respectively. A sample of three schools is chosen at random *with replacement* from the five schools. Let $\bar{X}$ denote the mean number of full-time Mathematics teachers per school in the sample. Find the sampling distribution of $\bar{X}$, and calculate its mean and standard deviation.

2 A pack consists of five cards numbered 1, 2, 3, 4, 5 respectively. Three cards are chosen at random with replacement from the pack. Let M denote the median of the numbers on the chosen cards. Find the sampling distribution of M and determine its mean.

3 A population consists of the five numbers 1, 2, 3, 4, 5. Find the population mean μ and variance σ^2.

Two numbers are chosen at random with replacement from the population. Let V denote the variance of the two chosen numbers. Find the sampling distribution of V and show that $E[V] \neq \sigma^2$.

4 Find the mean and standard error of the mean of a random sample of 10 values of a random variable having a Poisson distribution with mean 1.6.

5 A random variable X has a probability density function f given by
$$f(x) = 2x, \qquad 0 \leqslant x \leqslant 1,$$
$$f(x) = 0, \qquad \text{otherwise.}$$
Find the mean and variance of X.

Determine the mean and standard error of the mean of a random sample of five observations of X.

5.3 Distribution of the mean of a random sample from a normal distribution

Let $X_1, X_2, \ldots, X_n$ be a random sample of size n from the distribution of X, where $X \sim N(\mu, \sigma^2)$. Since the sample mean $\bar{X}$ is a linear combination of independent random variables $X_1, X_2, \ldots, X_n$, it is itself normally distributed (see Chapter 4, page 50), and from the results established above it follows that

$$\bar{X} \sim N\left(\mu, \frac{\sigma^2}{n}\right).$$

Example 4

The operational lifetimes of a particular brand of light bulbs are normally distributed with mean 1100 hours and standard deviation 20 hours.

a   Calculate the probability that the mean of the lifetimes of a random sample of 16 bulbs will be less than 1090 hours.

b   Find the smallest value of n for which there is a probability of at least 0.95 that the mean of the lifetimes of a random sample of n bulbs will be greater than 1095 hours.

Let X denote the operational lifetime of a bulb, then

$$X \sim N(1100, 20^2) \Rightarrow X \sim N(1100, 400).$$

a   Let $\bar{X}$ denote the mean of the lifetimes of 16 bulbs in the sample, then

$$\bar{X} \sim N\left(1100, \frac{400}{16}\right) \Rightarrow \bar{X} \sim N(1100, 25)$$

$$Z = \frac{\bar{X} - 1100}{5} \sim N(0, 1)$$

$$P(\bar{X} < 1090) = P\left(Z < \frac{1090 - 1100}{5}\right)$$

$$= P(Z < -2)$$

$$= 1 - 0.9772$$

$$= 0.0228$$

b   Let $\bar{X}$ denote the mean of the lifetimes of n bulbs in the sample, then

$$\bar{X} \sim N\left(1100, \frac{400}{n}\right)$$

$$Z = \frac{\bar{X} - 1100}{20/\sqrt{n}} \sim N(0, 1)$$

63

$$P(\bar{X} > 1095) > 0.95$$

$$P\left(Z > \frac{1095 - 1100}{20/\sqrt{n}}\right) > 0.95$$

$$\frac{-5}{20/\sqrt{n}} < -1.645$$

$$\frac{\sqrt{n}}{4} > 1.645$$

$$\sqrt{n} > 6.58$$

$$n > 43.2964$$

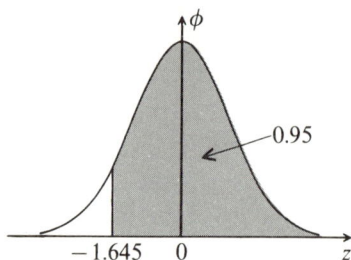

The required value of n is 44.

Exercise 5.3

1 A random sample of size 25 is taken from the distribution of X which is N(45, 16). Find the probability that the sample mean

    a will be greater than 46.5   **b** will lie between 44 and 47.

2 A random sample of size 10 is taken from the distribution of X which is N(150, 100). Find the probability that the sample mean

    a will lie between 148 and 155   **b** will be less than 160.

3 A random sample of size n is taken from a normal distribution having a standard deviation of 4. Find the least value of n for which the probability that the sample mean lies within 0.4 of the population mean exceeds 0.95.

4 When a particular instrument is used to measure lengths, the errors made are normally distributed with mean zero and standard deviation 0.5 mm. If the instrument is used n times to measure the length of a certain object, find the least value of n such that the probability that the mean error will be numerically less than 0.1 mm is greater than 0.9.

5 In a certain country, men have heights which are normally distributed with mean 172 cm and standard deviation 10 cm. Find the probability that a randomly chosen man has a height of less than 180 cm.

    A random sample of 4 men is chosen from the population. Find

    a the probability that all four men each have a height of less than 180 cm

    b the probability that the mean height of the four men is less than 180 cm.

5.4 Central Limit Theorem

If $X_1, X_2, \ldots, X_n$ is a random sample of size n from any distribution having mean μ and variance σ^2 then, for large n, the distribution of the sample mean $\bar{X}$ is approximately normal with mean μ and variance $\dfrac{\sigma^2}{n}$.

The truth of the above theorem will be assumed as its proof requires mathematics beyond A level. It is not possible to state a minimum value of n for which the approximation may be regarded as good, as this value depends upon the population distribution. A sample of 40 or more is considered to be a reasonable guideline for most populations, but the approximation can be quite good for smaller sample sizes, especially when the population distribution is symmetrical or nearly so.

The following useful result may be deduced from the Central Limit Theorem:

> If $X_1, X_2, \ldots, X_n$ is a random sample of size n from any distribution with mean μ and variance σ^2 then, for large n, the distribution of the sum of the sample values is approximately normal with mean $n\mu$ and variance $n\sigma^2$.

Proof

Let $T = X_1 + X_2 + \cdots + X_n$, then $T = n\bar{X}$.

Since $\bar{X}$ is approximately normal and T is a linear function of $\bar{X}$, then T is also approximately normal.

$$\begin{aligned} E[T] &= E[n\bar{X}] & V[T] &= V[n\bar{X}] \\ &= nE[\bar{X}] & &= n^2 V[\bar{X}] \\ &= n\mu & &= \frac{n^2 \sigma^2}{n} \\ & & &= n\sigma^2 \end{aligned}$$

$$T \text{ is approximately } N(n\mu, n\sigma^2)$$

Example 5

In manuscripts known to be written by author A, the mean and standard deviation of the number of words per sentence are 24 and 5 respectively. Find an approximate value for the probability that the mean number of words per sentence in a random sample of 100 sentences written by author A is greater than or equal to 25.

A long manuscript, which had been attributed to author A, was examined and a random sample of 100 sentences was found to contain a total of 2500 words. Comment on the attribution.

Let X denote the number of words per sentence in manuscripts written by A. The distribution of X is not known but its mean and variance are given by

$$\mu = 24 \quad \text{and} \quad \sigma^2 = 25.$$

Let $\bar{X}$ denote the mean number of words per sentence in a random sample of 100 sentences written by A. By the Central Limit Theorem the distribution of $\bar{X}$ is approximately normal with mean 24 and variance $\dfrac{25}{100}$.

Thus $Z = \dfrac{\bar{X} - 24}{\sqrt{0.25}}$ is approximately N(0, 1).

$$P(\bar{X} \geqslant 25) \simeq P\left(Z \geqslant \frac{25 - 24}{\sqrt{0.25}} \right)$$

$$= P(Z \geqslant 2)$$

$$= 0.023 \quad (3 \text{ d.p.})$$

This suggests that the attribution of the authorship of the manuscript to A is unlikely to be true.

Example 6

Independently for each day, the number of emergency admissions to a hospital has a Poisson distribution with mean 10. Find an approximate value for the probability that the total number of emergency admissions for a period of 13 weeks will exceed 950.

Let X_i denote the number of emergency admissions on the ith day of the period. Then X_i has a Poisson distribution with mean 10 and variance 10. The total number X of emergency admissions in the 13-week (91-day) period is given by

$$X = X_1 + X_2 + \cdots + X_{91}$$

Using the result deduced above from the Central Limit Theorem, the distribution of X is approximately normal with mean = 910 and variance 910.

Thus $Z = \dfrac{X - 910}{\sqrt{910}}$ is approximately N(0, 1).

$$P(X > 950) \simeq P\left(Z > \frac{950.5 - 910}{\sqrt{910}} \right)$$

In the above calculation a continuity correction is used because the distribution of X, a discrete random variable, is being approximated by a continuous normal distribution.

$$P(X > 950) \simeq P(Z > 1.343)$$

$$= 1 - 0.9104$$

$$= 0.0896$$

Exercise 5.4

1 A random sample of size 100 is taken from the binomial distribution B(20, 0.4). Find approximately the probability that the sample mean will be less than 7.6.

2 The masses of sacks of coal are distributed with mean 25.1 kg and standard deviation 0.4 kg. Find approximately the probability that a random sample of 50 of these sacks will have a mean mass of less than 25 kg.

3 The numbers of errors made by a typist on each page of typing are independent Poisson variables, each having a mean of 2. Calculate an approximate value for the probability that a book of 400 pages will contain less than 850 typing errors.

4 The times spent by a doctor examining patients in a clinic are uniformly distributed between 3 and 6 minutes. The doctor is to examine 40 patients in a morning session starting at 9.00 a.m. Find an approximate value for the probability that the session will finish before 12.15 p.m.

5 The probability density function of a random variable X is given by
$$f(x) = 2x, \qquad 0 \leqslant x \leqslant 1,$$
$$f(x) = 0, \qquad \text{otherwise.}$$

Find the mean and variance of X.

$\bar{X}$ is the mean of a random sample of 50 values of X. Find $P(\bar{X} > 0.6)$.

Miscellaneous Exercise ⬛ 5 ⬛

1 Find the mean μ and the variance σ^2 of the five numbers 0, 3, 3, 6, 6. A sample of three of these numbers is to be drawn at random *without* replacement. By making a list of all such samples, or otherwise, show that the sampling distribution of the sample mean $\bar{X}$ is given by the following table.

$\bar{x}$	2	3	4	5
$P(\bar{X} = \bar{x})$	0.1	0.4	0.3	0.2

Verify that $\bar{X}$ is an unbiased estimator of μ and calculate its variance.

If, instead, the sample is to be taken *with* replacement, state the value of the variance of the sample mean. *(JMB)*

2 One number is drawn at random from the set $\{1, 2, 3\}$. Denoting the number drawn by X, find the mean μ and variance σ^2 of X.

Samples of three numbers are to be drawn at random *with replacement* from the above set. By making a list of all such samples, or otherwise, find the sampling distributions of the sample mean $\bar{X}$ and the sample median M.

(i) Verify that $\bar{X}$ is an unbiased estimator of μ and calculate its variance.

(ii) Show that M is an unbiased estimator of μ and calculate its variance.

(iii) State, with a reason, which of $\bar{X}$ and M you consider to be the better estimator. *(JMB)*

STATISTICS 2

3 The number of days that each of five employees (A, B, C, D, E) in an office was absent from work during a year is shown in the following table:

Employee	A	B	C	D	E
No. of days absent	10	6	0	4	0

   a Calculate the mean μ and the variance σ^2 of the numbers of days these employees were absent from work.

   b Three of these employees are selected at random without replacement. Let $\bar{X}$ denote the mean number of days absent for the chosen three employees.

     (i) Determine the sampling distribution of $\bar{X}$ and display it in a table.

     (ii) Determine whether or not $\bar{X}$ is an unbiased estimator of μ.

     (iii) Find the variance of $\bar{X}$ and verify that it is equal to one-half of the variance of the sample mean if the three employees are chosen with replacement.

                                             (WJEC)

4 One number is to be drawn from the set $\{1, 2, 3, 4\}$. Denoting the drawn number by X, calculate the mean μ and the variance σ^2 of X.

Three numbers are to be drawn at random *with replacement* from the above set. Let $\bar{X}$ denote the mean and M the median of the three numbers drawn. (The median is the middle number when the three numbers are ranked according to size.)

   (i) Write down the values of the mean and the variance of $\bar{X}$.

   (ii) Show that $P(M = 4) = \dfrac{5}{32}$ and calculate the values of $P(M = r)$, for $r = 1, 2, 3$. Hence verify that M is an unbiased estimator of μ and calculate its variance.

                                             (JMB)

5 In a certain locality there are ten shops that sell a particular brand of honey. In three of the shops the honey is sold at 19p per jar, in six of the shops it is sold at 20p per jar, and in the remaining shop it is sold at 21p per jar. Calculate the mean and the standard deviation of the price per jar in the ten shops.

Suppose that three of these shops are selected at random (without replacement). Let M denote the mean price per jar for the three shops in the sample. Display the probability distribution of M in a table showing the various possible values of M and their respective probabilities. Using this table

   (i) verify that M is an unbiased estimator of the mean price per jar for all ten shops,

   (ii) calculate the standard deviation of M.

                                             (JMB)

6 The probability density function f of a continuous random variable X is given by

$$f(x) = k \sin x, \qquad 0 \leqslant x \leqslant \pi,$$
$$f(x) = 0, \qquad \text{otherwise.}$$

(i)   Show that $k = \dfrac{1}{2}$.

(ii) Find the mean of X.

(iii) Show that the variance of X is $\dfrac{\pi^2}{4} - 2$.

(iv) Denoting by $\bar{X}$ the mean of a random sample of 100 values of X, find, to three decimal places, an approximate value for

$$P\left(\bar{X} < \frac{27\pi}{50}\right).$$
(JMB)

7 Let $\bar{X}$ denote the mean of a random sample of 80 observations of the random variable X, whose probability density function f is given by

$$f(x) = \frac{3x(2 - x)}{4}, \qquad 0 < x < 2,$$
$$f(x) = 0, \qquad \text{otherwise.}$$

Using an appropriate approximation to the sampling distribution of $\bar{X}$, calculate the probabilities that $\bar{X}$ will be

(i)   less than 1,   (ii)   less than 0.95.   (*JMB*)

8 When a number is rounded to its nearest integer value the rounding error is defined to be the number minus its rounded value. When numbers are rounded to their nearest integer values the rounding errors may be regarded as independent observations from a uniform distribution over the interval $(-0.5, 0.5)$. Given that 75 numbers are rounded to their nearest integer values, use an appropriate distributional approximation to find the probability that the mean of the 75 rounding errors will be less than 0.05 in magnitude.   (*JMB*)

9 The continuous random variable X is distributed with probability density function f, where

$$f(x) = \frac{1}{2}, \qquad 0 \leqslant x \leqslant 1,$$
$$f(x) = \frac{(3 - x)}{4}, \qquad 1 < x \leqslant 3,$$
$$f(x) = 0, \qquad \text{otherwise.}$$

(i)   Find the mean and the variance of X.

(ii) Find the expected value of the area of a rectangle having adjacent sides of lengths X cm and $(5 - X)$ cm.

(iii) Let $\bar{X}$ denote the mean of a random sample of 100 observations of X. Use a normal approximation to the sampling distribution of $\bar{X}$ to find, to three decimal places, the probability that $\bar{X}$ will be less than 1.   (*JMB*)

10 The score X that may be obtained in one play of a certain game is a discrete random variable whose probability distribution is shown in the following table.

x	0	1	2	3
$P(X = x)$	0.4	0.2	0.2	0.2

(i) Find the mean and the variance of X.

(ii) Let $\bar{X}$ denote the mean of the scores in 136 independent plays of the game. Write down the mean value of $\bar{X}$ and show that its standard deviation is 0.1. Using a normal approximation to the sampling distribution of $\bar{X}$, find, correct to three decimal places, the probability that $\bar{X}$ will be less than unity. (*WJEC*)

11 The operational lifetimes of certain electronic components are found to be normally distributed with mean 5200 hours and standard deviation 400 hours. Calculate, to three significant figures,

(i) the proportion of such components having lifetimes between 4500 and 5800 hours,

(ii) the 67th percentile of this distribution,

(iii) the probability that a random sample of 25 components will have a mean lifetime in excess of 5000 hours. (*JMB*)

12 In a certain examination with a very large entry, the percentage marks obtained by the male candidates were found to follow a normal distribution with a mean of 54 and a standard deviation of 16. Let $\bar{X}$ denote the mean of the percentage marks scored by a random sample of 4 male candidates. What is the sampling distribution of $\bar{X}$? Calculate the probability that $\bar{X}$ will exceed 70 and the value c such that there is a probability of 0.95 that $\bar{X}$ will be within c marks of the mean mark of 54.

In the same examination the percentage marks obtained by the female candidates were found to follow a normal distribution with a mean of 59 and a standard deviation of 20. Let $\bar{Y}$ denote the mean of the percentage marks scored by a random sample of 5 female candidates. What is the sampling distribution of $\bar{Y} - \bar{X}$? Calculate the probability that the value of $\bar{Y}$ will be greater than the value of $\bar{X}$. (*WJEC*)

13 The number of miles travelled per week by a motorist is distributed normally with a mean of 640 and a standard deviation of 50.

a Calculate the probabilities that in a week he will travel
   (i) more than 600 miles,   (ii) between 600 and 700 miles.

b Calculate the probability that the average number of miles travelled per week over a complete year of 52 weeks will exceed 650 miles.

c If the car's petrol consumption is 30 miles per gallon, calculate the probability that the motorist will use less than 80 gallons of petrol over a period of 4 weeks. (*JMB*)

14 Oranges from a certain source have weights which are normally distributed with a mean of 150 g and a standard deviation of 4 g.

(i) Find the weight, to the nearest gram, which is exceeded by 90 per cent of the oranges.

(ii) Find the probability that a random sample of 4 oranges will have a mean weight in excess of 148 g.

When oranges are immersed in water their skins absorb some water and it may then be assumed that their weights are still normally distributed with a standard deviation of 4 g but that their weight will have increased to a new value μ g.

(iii) Given that 96.6% of oranges that have been immersed in water have weights in excess of 172.7 g, show that $\mu = 180$.

(iv) Find the probability that the combined weight of 4 oranges that have been immersed in water will exceed the combined weight of 5 oranges which have not been immersed in water. *(WJEC)*

15 When a number is rounded off to its nearest integer value, the rounding off error X may be regarded as a random variable which is uniformly distributed over the interval $\left(-\frac{1}{2}, \frac{1}{2}\right)$. Find the variance of X.

If n numbers are rounded off to their nearest integer values and then summed, the total error in the sum is given by
$$Y = X_1 + X_2 + \cdots + X_n,$$
where $X_1, X_2, \ldots, X_n$ are the individual errors in the n numbers, which may be assumed to be independent and uniformly distributed over $\left(-\frac{1}{2}, \frac{1}{2}\right)$. Write down the mean and the variance of Y in terms of n.

Suppose n is large enough to justify approximating the distribution of Y by a normal distribution. Show that there is a probability in excess of 0.9 that Y will be numerically less than $\frac{\sqrt{n}}{2}$. Find the largest integer value of n for which there is a probability of at least 0.6 that Y will be numerically less than 1.0. *(WJEC)*

Chapter 6

Estimation

In Chapter 5 the idea of using a statistic as an estimator of a population parameter was introduced. This idea will now be developed and discussed in more detail. Sample values may be used to calculate a single value (called a *point estimate*) or a range of values (called an *interval estimate*) to estimate an unknown population parameter. In this chapter only point estimates will be studied.

6.1 Unbiased estimators and estimates

Let θ denote an unknown parameter of the distribution of a random variable X and let T be a statistic derived from $\{X_1, X_2, \ldots, X_n\}$, a random sample of n observations taken from the distribution of X. T is said to be an *unbiased estimator* of θ if and only if

$$E[T] = \theta.$$

The value of T calculated from the observed sample values $x_1, x_2, \ldots, x_n$ is called an *unbiased estimate* of θ.

Notation

There are two commonly used notations for estimators and estimates. When a population parameter is denoted by a lower-case Greek letter, e.g. α, the estimator of this parameter is often denoted by the equivalent upper-case italic letter, A, and the corresponding estimate by the lower-case italic letter, a. Alternatively, when the population parameter is denoted by a lower-case italic letter, e.g. p, the estimator may be denoted by the symbol $\hat{P}$ and the corresponding estimate by $\hat{p}$.

The standard deviation of the sampling distribution of a statistic T is often referred to as the *standard error* of T, abbreviated to SE[T].

Relative efficiency and consistency

If T_1 and T_2 are two unbiased estimators of a population parameter θ such that SE[T_1] < SE[T_2], then T_1 is a better estimator of θ than T_2, since it is the more likely to give an estimate close to θ.

The ratio of $\dfrac{V[T_1]}{V[T_2]}$ is called the *relative efficiency* of T_1 with respect to T_2, it is often expressed as a percentage.

If T is an estimator of θ derived from a random sample of size n and as $n \to \infty$, $E[T] \to \theta$ and $SE[T] \to 0$, then T is said to be a *consistent estimator* of θ.

Point estimator of a population mean

The most commonly used point estimator for a population mean is the mean $\bar{X}$ of a random sample of n observations taken from the population.

In Chapter 5, page 61, it was shown that $E[\bar{X}] = \mu$, so that $\bar{X}$ is an unbiased estimator for μ. The value, $\bar{x}$, of $\bar{X}$ calculated from the observed sample values $x_1, x_2, \ldots, x_n$ is an unbiased estimate of μ.

It was also shown that $V[\bar{X}] = \dfrac{\sigma^2}{n}$, so that as $n \to \infty$, $SE[\bar{X}] \to 0$. Thus $\bar{X}$ is a consistent estimator of μ.

Example 1

$\{X_1, X_2, X_3\}$ is a random sample taken from a population distribution with mean μ and variance σ^2; $T_1 = \dfrac{(X_1 + 2X_2 + 3X_3)}{6}$ and $T_2 = X_1 + X_2 - X_3$. Show that T_1 and T_2 are both unbiased estimators of μ and determine which is the better.

$$E[T_1] = E\left[\frac{(X_1 + 2X_2 + 3X_3)}{6}\right]$$
$$= \frac{(E[X_1] + 2E[X_2] + 3E[X_3])}{6}$$
$$= \frac{(\mu + 2\mu + 3\mu)}{6}$$
$$= \mu$$

Therefore T_1 is an unbiased estimator of μ.

$$E[T_2] = E[X_1 + X_2 - X_3]$$
$$= E[X_1] + E[X_2] - E[X_3]$$
$$= \mu + \mu - \mu$$
$$= \mu$$

Therefore T_2 is an unbiased estimator of μ.

$$V[T_1] = V\left[\frac{(X_1 + 2X_2 + 3X_3)}{6}\right]$$

$$= \frac{V[X_1 + 2X_2 + 3X_3]}{36}$$

$$= \frac{(V[X_1] + 4V[X_2] + 9V[X_3])}{36}$$

$$= \frac{(\sigma^2 + 4\sigma^2 + 9\sigma^2)}{36}$$

$$= \frac{14\sigma^2}{36}$$

$$V[T_2] = V[X_1 + X_2 - X_3]$$
$$= V[X_1] + V[X_2] + V[X_3]$$
$$= \sigma^2 + \sigma^2 + \sigma^2$$
$$= 3\sigma^2$$

$V[T_2] > V[T_1]$ therefore T_1 is the better (or more efficient) estimator.

Exercise 6.1

1 $\{X_1, X_2, X_3\}$ is a random sample taken from a population with mean μ and variance σ^2. Find which of the following estimators for μ are unbiased.

$$T_1 = \frac{(X_1 + 2X_2 + X_3)}{4} \qquad T_2 = 3X_1 - X_2 - X_3$$

$$T_3 = \frac{(X_1 + 3X_2 + 2X_3)}{5} \qquad T_4 = \frac{(X_1 + X_2 + X_3)}{3}$$

2 Determine which of the estimators in Question **1** is the most efficient.

3 $\{X_1, X_2, X_3, X_4\}$ is a random sample taken from a population with mean μ and variance σ^2; $T_1 = \frac{(X_1 + X_2 + X_3 + X_4)}{4}$ and $T_2 = \frac{(X_1 + X_2 + X_3)}{3}$.
Show that T_1 and T_2 are both unbiased estimators for μ and determine which is the more efficient.

4 $\{X_1, X_2\}$ is a random sample from a population with mean μ and variance σ^2. The statistic $T = aX_1 + bX_2$ is to be used as an estimator for μ.
  a Find $E[T]$ and $V[T]$ in terms of μ, σ^2, a and b.
  b Show that, for T to be an unbiased estimator for μ, $a + b = 1$.
  c Hence find the values of a and b for $V[T]$ to be a minimum.

5 A random observation X is taken from a population with mean μ and variance 4; an independent random observation Y is taken from a population with mean 2μ and variance 8. Given that $T = aX + bY$ is an unbiased estimator of μ, find the values of a and b such that T has a minimum variance.

6.2 Point estimator for a population proportion

Suppose that a population has an unknown proportion p having a particular quality and that, in a random sample of n observations taken from the population, X have that quality, then $X \sim B(n, p)$.

Therefore $\qquad E[X] = np \qquad$ and $\qquad V[X] = np(1 - p).$

Intuitively, the sample proportion $\hat{P} = \dfrac{X}{n}$ should be a sensible estimator for the population proportion p.

$$E[\hat{P}] = E\left[\frac{X}{n}\right] \qquad\qquad V[\hat{P}] = V\left[\frac{X}{n}\right]$$

$$E[\hat{P}] = \frac{1}{n}E[X] \qquad\qquad V[\hat{P}] = \frac{1}{n^2}V[X]$$

$$E[\hat{P}] = \frac{1}{n}np \qquad\qquad V[\hat{P}] = \frac{1}{n^2}np(1 - p)$$

$$E[\hat{P}] = p \qquad\qquad V[\hat{P}] = \frac{p(1 - p)}{n}$$

Thus the sample proportion $\hat{P}$ is an unbiased estimator for the unknown population proportion p. The standard error of $\hat{P}$ is given by

$$SE[\hat{P}] = \sqrt{\frac{p(1 - p)}{n}}$$

and since $SE[\hat{P}] \to 0$ as $n \to \infty$, $\hat{P}$ is a consistent estimator for p.

As $SE[\hat{P}]$ is a function of the unknown quantity p, it cannot be evaluated even after the value of the sample proportion has been calculated. However, an upper limit for $SE[\hat{P}]$ may be calculated. By differentiating, or by completing the square, it is easy to show that the maximum value of $p(1 - p)$ is $\dfrac{1}{4}$, so that, for any value of p,

$$SE[\hat{P}] \leqslant \sqrt{\frac{1}{4n}}.$$

In later work it will be necessary to estimate the standard error of $\hat{P}$. One way of doing this is to replace p in the formula for $SE[\hat{P}]$ by an unbiased estimate, the observed value of the sample proportion $\hat{p} = \dfrac{x}{n}$, giving

$$ESE[\hat{P}] = \sqrt{\frac{\hat{p}(1 - \hat{p})}{n}}$$

where ESE stands for 'estimated standard error'. It may be noted that this $ESE[\hat{P}]$ is, in fact, a biased estimator of $SE[\hat{P}]$.

The results established above assume that sampling with replacement was carried out, but the results are approximately true when the sampling without replacement is performed, provided that the size of the population is very large in comparison with the sample size.

Example 2

In an opinion poll conducted in a certain constituency it was found that 216 of a random sample of 400 voters stated that they intended to vote for party A in a forthcoming election.

Calculate an unbiased estimate for the proportion of the electorate in the constituency that will vote for party A in the election. Calculate an upper bound and an approximate value of the standard error of this estimate.

It may be assumed that the size of the electorate is very large so that the results established above may be applied.

An unbiased estimate for proportion p of all voters who vote for party A is $\hat{p} = \dfrac{216}{400} = 0.54.$

The upper bound is given by $\sqrt{\left(\dfrac{1}{4n}\right)} = \sqrt{\left(\dfrac{1}{1600}\right)} = 0.025.$

The standard error of $\hat{P}$ is given by $\text{SE}[\hat{P}] = \sqrt{\dfrac{p(1-p)}{n}}.$

Replacing p with 0.54 gives $\quad \text{ESE}[\hat{P}] = \sqrt{\dfrac{0.54 \times 0.46}{400}} = 0.0249 \quad$ (3 s.f.).

It has been assumed that sampled voters answered independently and truthfully, and that none will change their mind on voting day.

Exercise 6.2

1   In an opinion poll 350 of a random sample of 1000 voters said that they will vote for party B in the next election. Calculate an unbiased estimate for the proportion of the electorate that will vote for party B in the election. Calculate an upper bound and an approximate value for the standard error of this estimate.

2   In a random sample of 50 pupils of a large school 12 were found to be left-handed. Calculate an unbiased estimate for the proportion of pupils in the school that are left-handed. Calculate an approximate value for the standard error of this estimate and determine an upper bound for the standard error.

3   A seed merchant produces large numbers of seeds of a particular variety. Of a random sample of 200 seeds planted 184 germinated. Find an unbiased estimate for the proportion of the seeds produced that will germinate and also find an approximate value for the standard error of this estimate. Determine an upper bound for this standard error.

4 A random sample of 300 cat owners were asked about the brand of cat food they bought for their pets; 252 said they bought brand X. In a second random sample of 200 a total of 150 said they bought brand X. Calculate three unbiased estimates for the proportion of all cat owners who buy brand X. Indicate which of your three estimates has the smallest standard error.

5 An inspector selects a random sample of size 30 from a batch of manufactured components and he discovers that X_1 are defective. A second inspector selects independently a random sample of 20 components and discovers that X_2 are defective. Show that

$$\hat{P}_1 = \frac{1}{2}\left(\frac{X_1}{30} + \frac{X_2}{20}\right) \text{ and } \hat{P}_2 = \frac{X_1 + X_2}{50}$$

are both unbiased estimators for the proportion of defective components in the batch. Determine which is the more efficient.

6.3 Point estimator for a population variance

Suppose that $\{X_1, X_2, \ldots, X_n\}$ is a random sample of size n taken from the distribution of X with unknown mean μ and unknown variance σ^2. Intuitively, the sample variance, given by

$$V = \frac{\sum_{i=1}^{n} (X_i - \bar{X})^2}{n},$$

would seem to be a sensible estimator for σ^2, but, as shown below, it has the disadvantage that it is a biased estimator.

To show that V is a biased estimator for σ^2, $E[V]$ is evaluated.

$$E[V] = E\left[\frac{\sum_{i=1}^{n} (X_i - \bar{X})^2}{n}\right].$$

Using a result established on page 24 of Book 1.

$$E[V] = E\left[\frac{\sum_{i=1}^{n} X_i^2 - n\bar{X}^2}{n}\right].$$

But the expected value of a linear function of random variables is equal to the linear function of the expected values of the random variables (see the first result stated on page 37).

$$E[V] = \frac{1}{n}\left[\sum_{i=1}^{n} E[X_i^2] - nE[\bar{X}^2]\right] \tag{1}$$

But X_i has the same distribution as X, therefore

$$E[X_i^2] = V[X_i] + \{E[X_i]\}^2$$

i.e. $E[X_i^2] = \sigma^2 + \mu^2$

and since the sampling distribution of $\bar{X}$ has mean μ and variance $\dfrac{\sigma^2}{n}$

$$E[\bar{X}^2] = V[\bar{X}] + \{E[\bar{X}]\}^2$$

i.e. $\quad E[\bar{X}^2] = \dfrac{\sigma^2}{n} + \mu^2.$

Substituting these results in (1)

$$E[V] = \frac{1}{n}\left[\sum_{i=1}^{n}(\sigma^2 + \mu^2) - n\left(\frac{\sigma^2}{n} + \mu^2\right)\right]$$

$$= \frac{1}{n}[n(\sigma^2 + \mu^2) - \sigma^2 - n\mu^2]$$

$$= \frac{(n-1)\sigma^2}{n}.$$

Since this is not σ^2, V is a biased estimator for σ^2.

Now consider the estimator $\dfrac{nV}{(n-1)}$.

$$E\left[\frac{nV}{(n-1)}\right] = \frac{n}{(n-1)}E[V]$$

$$= \frac{n}{(n-1)} \cdot \frac{(n-1)\sigma^2}{n}$$

$$= \sigma^2$$

Thus $\dfrac{nV}{(n-1)}$ is an unbiased estimator for σ^2 and it will be denoted by S^2.

$$S^2 = \frac{\sum_{i=1}^{n}(X_i - \bar{X})^2}{(n-1)}$$

is called the *sample unbiased estimator* for the population variance σ^2.

The corresponding unbiased estimate is denoted by s^2 and is usually evaluated using the formula

$$s^2 = \frac{\sum_{i=1}^{n} x_i^2 - n\bar{x}^2}{(n-1)}.$$

Example 3

Ten 1 kg packets of sugar of a particular brand were chosen at random and their masses were measured to the nearest g. The results in kg were as follows

| 1.012 | 1.015 | 1.017 | 0.998 | 1.011 |
| 1.021 | 0.996 | 1.016 | 1.013 | 1.011. |

Calculate unbiased estimates for the mean μ and the variance σ^2 of the masses of all 1 kg packets of the brand.

Let the observed masses be $x_1, x_2, \ldots, x_{10}$.
Then $\sum x_i = 10.11$ and $\sum x_i^2 = 10.221786$.

Unbiased estimate for μ is given by $\bar{x} = \dfrac{\sum x_i}{n}$

$$\bar{x} = \frac{10.11}{10}$$

$$\bar{x} = 1.011.$$

Unbiased estimate for σ^2 is given by $s^2 = \dfrac{\sum x_i^2 - n\bar{x}^2}{n - 1}$

$$s^2 = \frac{10.221786 - 10 \times 1.011^2}{9}$$

$$s^2 = 0.000064.$$

Exercise 6.3

1 A random sample of eight observations from a distribution gave values

$$2.6, \quad 2.3, \quad 3.1, \quad 2.8, \quad 2.6, \quad 2.1, \quad 2.9, \quad 3.2.$$

Find unbiased estimates for the mean and the variance of the distribution.

2 A random sample of six observations from a distribution gave values

$$33.2, \quad 36.6, \quad 35.7, \quad 36.1, \quad 34.3, \quad 35.9.$$

Find unbiased estimates for the mean and the variance of the distribution.

3 The height x cm of each man in a random sample of 200 men living in the U.K. was measured. The following results were obtained.

$$\sum x = 35250, \qquad \sum x^2 = 6233409.$$

Calculate unbiased estimates for the mean and the variance of the heights of men living in the U.K.

4 The mass x kg of potatoes produced by each plant in a random sample of 250 plants grown by a certain method was measured. The following results were obtained:

$$\sum x = 3660, \qquad \sum x^2 = 54379.$$

Calculate, to two decimal places, unbiased estimates for the mean and the variance of the masses of potatoes produced per plant by the method.

5 A random sample of 100 observations taken from a Poisson distribution with unknown mean α are such that the sum of the observations is 211.1 and the sum of their squares is 659.32. Calculate two unbiased estimates for α.

6 The distribution of a random variable X has unknown mean μ and variance σ^2. From a random sample of n_1 values of X unbiased estimates $\bar{x}_1$ and $s_1{}^2$ for μ and σ^2 respectively were calculated; from a second random sample of size n_2 corresponding unbiased estimates $\bar{x}_2$ and $s_2{}^2$ were calculated. Show that

$$\frac{n_1\bar{x}_1 + n_2\bar{x}_2}{n_1 + n_2}, \qquad \frac{(n_1 - 1)s_1{}^2 + (n_2 - 1)s_2{}^2}{n_1 + n_2 - 2}$$

are unbiased estimates for μ and σ^2 respectively.

STATISTICS 2

6.4 Further examples

Example 4

A continuous random variable X is uniformly distributed between 0 and θ, where θ is an unknown constant.

a Write down the mean and the variance of X in terms of θ.

b $\bar{X}$ is the mean of a random sample of n observations of X. Show that $T = 2\bar{X}$ is an unbiased and consistent estimator for θ.

c Calculate an unbiased estimate for θ from the following random sample

1.6, 1.8, 2.1, 4.4, 1.8, 2.0, 1.9, 1.7, 2.0, 2.0.

Explain why the estimate obtained is unreasonable.

a Applying standard results derived on page 138 of Book 1

$$E[X] = \frac{\theta}{2} \qquad V[X] = \frac{\theta^2}{12}$$

b
$$E[T] = E[2\bar{X}]$$
$$= 2E[\bar{X}]$$

But $E[\bar{X}] = E[X]$ therefore
$$E[T] = 2E[X]$$
$$= 2 \times \frac{\theta}{2}$$
$$= \theta.$$

Therefore T is an unbiased estimator for θ.
$$V[T] = V[2\bar{X}]$$
$$= 4V[\bar{X}]$$

But $V[\bar{X}] = \frac{V[X]}{n}$ therefore
$$V[T] = 4\frac{V[X]}{n}$$
$$= 4 \times \frac{\theta^2}{12} \times \frac{1}{n}$$
$$= \frac{\theta^2}{3n}.$$

As $n \to \infty$, $V[T] \to 0$, therefore T is a consistent estimator for θ.

c For the sample values $\sum x = 21.3$ $\qquad \therefore \bar{x} = 2.13.$

Therefore unbiased estimate for θ is $2 \times 2.13 = 4.16$.

This estimate is unreasonable since it is known that θ is at least 4.4, the largest value in the sample.

ESTIMATION

Example 5

The continuous random variable X has probability density function f
given by

$$f(x) = \theta x + \frac{3}{2}x^2 \qquad\qquad -1 < x < 1,$$

$$f(x) = 0 \qquad\qquad\qquad \text{otherwise,}$$

where θ is an unknown constant.

a   Find, in terms of θ, $P(X > 0)$.

b   If Y denotes the number of positive values in a random sample of n
values of X, name the distribution of Y and write down $E[Y]$ and
$V[Y]$ in terms of θ.

c   Show that $T = \dfrac{2Y}{n} - 1$ is an unbiased estimator for θ and find its
variance.

a $P(X > 0) = \displaystyle\int_0^1 \left(\theta x + \frac{3}{2}x^2\right) dx$

$$= \left[\theta\frac{x^2}{2} + \frac{x^3}{2}\right]_0^1 = \frac{(\theta + 1)}{2}$$

b $Y \sim B\left(n, \dfrac{(\theta + 1)}{2}\right)$

For $B(n, p)$, mean $= np$ $\qquad\qquad E[Y] = n\dfrac{(\theta + 1)}{2}$

variance $= np(1 - p)$ $\qquad\qquad V[Y] = n\dfrac{(\theta + 1)}{2}\left(1 - \dfrac{\theta + 1}{2}\right).$

Simplifying $\qquad\qquad\qquad V[Y] = n\dfrac{(1 - \theta^2)}{4}.$

c $\qquad E[T] = E\left[2\dfrac{Y}{n} - 1\right]$

$$= \frac{2}{n}E[Y] - 1 = \frac{2}{n} \times n\frac{(\theta + 1)}{2} - 1 = \theta$$

Therefore T is an unbiased estimator for θ.

$$V[T] = V\left[2\frac{Y}{n} - 1\right]$$

$$= \frac{4}{n^2}V[Y] = \frac{4}{n^2} \times n\frac{(1 - \theta^2)}{4}$$

Therefore $\qquad V[T] = \dfrac{1 - \theta^2}{n}.$

81

Miscellaneous Exercise ⬛ 6 ⬛

1 A seed merchant sells trays, each containing 10 flower seeds in specially prepared soil. The number of seeds germinating in each of the 100 trays was noted and the following results were obtained.

No. of seeds germinating	5	6	7	8	9	10
No. of trays	1	12	27	33	23	4

   (i)   Find the median number of seeds germinating per tray in this sample.

   (ii) Find an unbiased estimate of the proportion of these seeds that germinate and also find, correct to three decimal places, an approximate value of the standard error of this estimate.   (*JMB*)

2 A random sample of nine observations is drawn from a population distribution having mean μ and variance σ^2. The sample mean is 6 and the sample unbiased estimate of σ^2 is 9. Given that an additional random observation from the distribution has the value 5, find unbiased estimates of μ and σ^2 based on all ten observations.   (*JMB*)

3 Explain what is meant by the sampling distribution of an estimator. Given that a random sample of n observations is drawn from a normal distribution having mean μ and variance σ^2, specify the sampling distribution of the sample mean. State what can be said about this sampling distribution when n is large and the parent population is *not* normal.

   Explain what you understand by

   (i)   an unbiased estimator,

   (ii) a consistent estimator,

   (iii) the relative efficiency of two estimators of the same parameter.

   In order to estimate the mean μ of a population, random observations x_1, x_2, x_3 are taken of a random variable X which has variance σ^2. Find the relative efficiency of the two estimators $\hat{\mu}_1$ and $\hat{\mu}_2$, where

$$\hat{\mu}_1 = \frac{x_1 + x_2 + x_3}{3} \qquad \hat{\mu}_2 = \frac{x_1 + 2x_2}{3}. \qquad (JMB)$$

4 The random variables X_1, X_2 and X_3 are independent. Their means are μ, 2μ and 3μ, respectively, and each has unit variance. Show that each of the statistics

$$T_1 = \frac{1}{6}(X_1 + X_2 + X_3) \qquad \text{and} \qquad T_2 = \frac{1}{3}\left(X_1 + \frac{1}{2}X_2 + \frac{1}{3}X_3\right)$$

   is an unbiased estimator of μ. Determine which of T_1 and T_2 is the better estimator of μ.   (*WJEC*)

5 The random variable X has the Poisson distribution

$$P(X = k) = e^{-\theta}\frac{\theta^k}{k!}, \qquad k = 0, 1, 2, \ldots .$$

Show that (i) $E[X] = \theta$,

(ii) $E[X^2 - X] = \theta^2$,

(iii) $\mathrm{Var}[X] = \theta$.

A random sample of 9 observations of X had values

$$8, \quad 6, \quad 4, \quad 7, \quad 5, \quad 3, \quad 1, \quad 6, \quad 5.$$

Calculate two unbiased estimates of θ and one unbiased estimate of θ^2 based on all 9 observations. *(WJEC)*

6 The continuous random variable X has probability density function f, where

$$f(x) = \frac{1}{2}(1 + \theta x), \qquad -1 < x < 1,$$

$$f(x) = 0, \qquad \text{otherwise,}$$

and θ is a constant whose value lies somewhere in the interval -1 to $+1$.
  (i) Determine, in terms of θ, the probability that a randomly observed value of X will be positive.
  (ii) Let R denote the number of positive values in a random sample of n observations of X. Name the distribution of R. Show that $T_1 = \dfrac{4R}{n} - 2$ is an unbiased estimator of θ and that the variance of T_1 is $\dfrac{(4 - \theta^2)}{n}$.
  (iii) Let $\bar{X}$ denote the mean of a random sample of n observations of X. Show that $T = 3\bar{X}$ is an unbiased estimator of θ and that its variance is $\dfrac{(3 - \theta^2)}{n}$.
  (iv) State, with your reason, which of T_1 and T_2 you would prefer for estimating θ. *(WJEC)*

7 A random variable X can take only the values 1, 2 and 3, the respective probabilities of these values being θ, θ and $1 - 2\theta$, where $0 < \theta < \frac{1}{2}$.

Determine the mean and the variance of X in terms of θ.

In a random sample of n values of X, let $\bar{X}$ denote the mean of the sample values, and let R denote the number of 3's among the sample values. Show that

$$\hat{P}_1 = 1 - \frac{\bar{X}}{3} \quad \text{and} \quad \hat{P}_2 = \frac{1}{2}\left(1 - \frac{R}{n}\right)$$

are both unbiased estimators of θ.

Determine which of $\hat{P}_1$ and $\hat{P}_2$ has the smaller standard error. *(WJEC)*

8 A discrete random variable X can take the values 0, 1 and 2 only, with respective probabilities $\frac{\theta}{2}$, $1 - \theta$ and $\frac{\theta}{2}$, where θ is an unknown number between 0 and 1. Let X_1 and X_2 denote two randomly observed values of X. List the possible values of $\{X_1, X_2\}$ that may arise and calculate the probability of each possibility; verify that your probabilities sum to unity.

By calculating the value $(X_1 - X_2)^2$ for each possible $\{X_1, X_2\}$ determine the sampling distribution of $(X_1 - X_2)^2$. Hence show that $Y = \frac{(X_1 - X_2)^2}{2}$ is an unbiased estimator of θ and express its sampling variance in terms of θ.

Since θ is the probability that X will not take the value 1, another possible estimator of θ is the proportion of sample values not equal to 1; for a sample of two observations this estimator is given by $Z = \frac{N}{2}$, where N is the number of observations (0, 1 or 2) not equal to 1. State, giving your reasons, which of Y and Z you would prefer as an estimator of θ.

(WJEC)

9 Derive the mean and the variance of the binomial distribution.

A drawing pin was tossed n times in succession and the number of occasions, X, that it alighted with its point upright was noted. Some time later, the same drawing pin was tossed a further m times and the number of occasions, Y, that it alighted with its point upright was noted. Show that

$$Z_1 = \frac{1}{2}\left(\frac{X}{n} + \frac{Y}{m}\right) \qquad \text{and} \qquad Z_2 = \frac{X + Y}{n + m}$$

are both unbiased estimators of p, the probability that when tossed the drawing pin will alight with its point upright. State, with reasons, which of these two estimators you would regard to be preferable to use. *(WJEC)*

10 A continuous random variable X is uniformly distributed over the interval $(a - c, a + c)$. Determine the expected values of X and X^2.

An instrument for measuring the length of a line is such that the recorded value for a line of length a cm is equally likely to be any value in the interval $(a - c, a + c)$, where c (less than a) is a known positive constant. The instrument is used to obtain two independent observations, X_1 and X_2, of the length of a side of a square. Consider the two following methods for estimating the area A of the square:

Method 1   Estimate A by $T_1 = X_1 X_2$,

Method 2   Estimate A by $T_2 = \dfrac{(X_1 + X_2)^2}{4}$.

Show that Method 1 is the only one of these two methods that will give an unbiased estimate of the area of the square, and determine the standard error of this estimate in terms of a and c. *(WJEC)*

Chapter 7

Confidence intervals

In the last chapter the properties of some unbiased estimators of population parameters were studied. In many examples, an estimate for a particular parameter was calculated; this point estimate had the disadvantage that there was no associated indication of the closeness of the estimate to the true value of the parameter. This disadvantage will be overcome in the present chapter by calculating an interval estimate for the population parameter together with a numerical measure of the confidence with which the result is asserted. For example, a *95% confidence interval* for an unknown population parameter is an interval calculated by a method which is such that there is a probability of 0.95 of obtaining an interval which actually contains the true value of the parameter. These ideas will be illuminated by the specific examples that follow.

7.1 Confidence intervals for the mean of a normal distribution with known variance

Let $\bar{X}$ be the mean of a random sample of n observations of X, which is normally distributed with mean μ (unknown) and variance σ^2 (known). The sampling distribution of $\bar{X}$ is also normal, with mean μ and standard error $SE[\bar{X}] = \dfrac{\sigma}{\sqrt{n}}$.

Hence $\quad Z = \dfrac{\bar{X} - \mu}{SE[\bar{X}]} \sim N(0, 1).$

From **Table 3** $\quad P(Z < 1.96) = 0.975.$

Therefore $\quad P(Z < -1.96) = 1 - 0.975 = 0.025.$

Combining $\quad P(-1.96 < Z < 1.96) = 0.95.$

Therefore $\quad P\left(-1.96 < \dfrac{\bar{X} - \mu}{SE[\bar{X}]} < 1.96\right) = 0.95.$

Rearranging $\quad P(-1.96 \times SE[\bar{X}] < \bar{X} - \mu < 1.96 \times SE[\bar{X}]) = 0.95$

$\quad\quad\quad\quad P(\bar{X} - 1.96 \times SE[\bar{X}] < \mu < \bar{X} + 1.96 \times SE[\bar{X}]) = 0.95. \quad (1)$

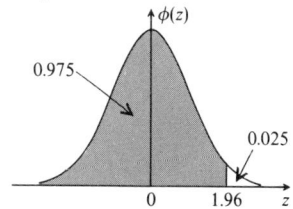

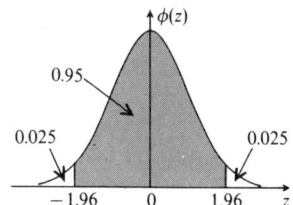

85

This appears to be a probabilistic statement about μ, but this is not so. Probabilistic statements cannot be made about constants (whether known or unknown), thus equation (1) is a probabilistic statement about $\bar{X}$.
An alternative version is as follows:

the random interval $(\bar{X} - 1.96 \times SE[\bar{X}], \bar{X} + 1.96 \times SE[\bar{X}])$ has a probability of 0.95 of including the true value of μ.

When the observed value of the sample mean is $\bar{x}$, the interval

$$(\bar{x} - 1.96 \times SE[\bar{X}], \bar{x} + 1.96 \times SE[\bar{X}])$$

is called a *95% confidence interval for μ* and the end points of the interval are called *95% confidence limits for μ*.

It should be noted that the particular 95% confidence interval for μ obtained may be one of the 95% of such intervals that contain μ or it may be one of the 5% of such intervals that do not contain μ. It is incorrect to say that μ lies within the observed interval $(\bar{x} - 1.96 \times SE[\bar{X}], \bar{x} + 1.96 \times SE[\bar{X}])$ with probability 0.95.

Example 1

The masses of bags of sugar filled by a certain machine are normally distributed with standard deviation 25 g. A random sample of 100 of these bags had a mean mass of 1011 g. Find a 95% confidence interval for μ, the mean mass of bags of sugar filled by the machine.

$\bar{X}$ is normally distributed with mean μ and $SE[\bar{X}] = \dfrac{25}{\sqrt{100}}$.

The 95% confidence limits for μ are given by

$$\bar{x} \pm 1.96 \times SE[\bar{X}]$$

Substituting $\quad 1011 \pm 1.96 \times \dfrac{25}{\sqrt{100}}$

Evaluating $\quad 1011 \pm 4.9$

The 95% confidence interval for μ is (1006.1, 1015.9).

Other confidence levels

The choice of 95% as confidence level in the work above was clearly arbitrary. Any confidence level may be used, but those most commonly used in practice are 90%, 95% and 99%. For a particular set of data, the higher the selected confidence level the greater is the width of the confidence interval. If the assumption that μ lies in the confidence interval when, in fact, it does not, has serious consequences then a high confidence level should be chosen.

As before, let $\bar{X}$ denote the mean of a random sample of n observations of X, which is normally distributed with mean μ (unknown) and variance σ^2 (known).

Then $Z = \dfrac{\bar{X} - \mu}{SE[\bar{X}]} \sim N(0, 1)$.

Let z_p be such that $P(Z > z_p) = p$, where p has a specific value between 0 and 0.5.

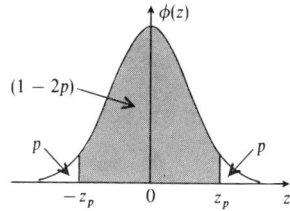

From the diagram $\quad P(-z_p < Z < z_p) = 1 - 2p,$

$$P\left(-z_p < \frac{\bar{X} - \mu}{SE[\bar{X}]} < z_p\right) = 1 - 2p,$$

Rearranging, as on page 85,
$$P(\bar{X} - z_p SE[\bar{X}] < \mu < \bar{X} + z_p SE[\bar{X}]) = 1 - 2p.$$

Therefore $100(1 - 2p)\%$ confidence limits for μ are given by
$$\bar{x} \pm z_p SE[\bar{X}]$$
where $\bar{x}$ is the observed value of the sample mean.

The interval $(\bar{x} - z_p SE[\bar{X}], \bar{x} + z_p SE[\bar{X}])$ is called a *symmetric* $100(1 - 2p)\%$ *confidence interval for* μ. It said to be symmetric because the point estimate $\bar{x}$ for μ is the mid-point of the interval.

Values of z_p for some values of p are given (to 2 d.p.) in **Table 3b**. The most commonly used values of z_p are given (to 3 d.p.) in the table below.

p	0.05	0.025	0.005
$100(1 - 2p)\%$	90%	95%	99%
z_p	1.645	1.960	2.576

Thus the most common confidence limits for μ are as follows:
90% confidence limits for μ $\quad\quad \bar{x} \pm 1.645 \times SE[\bar{X}]$
95% confidence limits for μ $\quad\quad \bar{x} \pm 1.960 \times SE[\bar{X}]$
99% confidence limits for μ $\quad\quad \bar{x} \pm 2.576 \times SE[\bar{X}]$.

Example 2

It is known that the readings on a chemical balance are normally distributed with mean equal to the true weight of the object and with standard deviation 0.5 mg.

a   Sixteen independent weighings of an object gave a mean reading of 15.6 mg. Calculate a symmetric 90% confidence interval for the true weight of the object.

b   Determine the smallest number of independent weighings of an object that should be made so that the 99% confidence interval for the true weight of the object has a width of less than 0.5 mg.

a $\bar{X}$, the mean of 16 independent weighings, is normally distributed with mean μ, the true weight, and standard error $\text{SE}[\bar{X}] = \dfrac{0.5}{\sqrt{16}}$

The 90% confidence limits are given by
$$\bar{x} \pm 1.645 \times \text{SE}[\bar{X}].$$

Substituting $15.6 \pm 1.645 \times \dfrac{0.5}{\sqrt{16}}.$

Evaluating $15.6 \pm 0.21.$

The 90% confidence interval is (15.39, 15.81).

b $\bar{X}$, the mean of n independent weighings, is normally distributed with mean μ and $\text{SE}[\bar{X}] = \dfrac{0.5}{\sqrt{n}}.$

The 99% confidence limits are given by
$$\bar{x} \pm 2.576 \times \text{SE}[\bar{X}].$$

The width w of this interval is $\left(\bar{x} + 2.576 \times \dfrac{0.5}{\sqrt{n}}\right) - \left(\bar{x} - 2.576 \times \dfrac{0.5}{\sqrt{n}}\right).$

Therefore $w = 2 \times 2.576 \times \dfrac{0.5}{\sqrt{n}}.$

For w to be less than 0.5, $2 \times 2.576 \times \dfrac{0.5}{\sqrt{n}} < 0.5.$

Rearranging $2 \times 2.576 \times \dfrac{0.5}{0.5} < \sqrt{n}.$

Squaring $26.543104 < n.$

Therefore the smallest value of n required is 27.

Exercise 7.1

1   The heights of women in a certain country are normally distributed with standard deviation 3 cm. A random sample of 25 of these women had a mean height of 162 cm. Calculate

a   95%
b   90%
c   99% confidence intervals for the mean height of women in the country.

2   The weights of male employees in a large firm are normally distributed with standard deviation 10 kg. A random sample of 16 of these men had a mean weight of 82.5 kg. Calculate

a   95%
b   90%
c   99% confidence intervals for the mean weight of male employees of the firm.

3 The masses of bags of flour produced by a certain company are normally distributed with standard deviation 0.1 kg.

  a A random sample of 100 bags had a mean mass of 1.6 kg. Find a 95% confidence interval for the mean mass of bags of flour produced by the company.

  b Find the least sample size such that a 95% confidence interval for the mean mass of flour bags has a width of less than 0.02 kg.

4 The lifetimes of light bulbs of brand A are normally distributed with standard deviation 33 hours.

  a A random sample of 9 light bulbs has a mean lifetime of 983 hours. Find a 90% confidence interval for the mean lifetime of light bulbs of brand A.

  b Find the least sample size such that a 99% confidence interval for the mean lifetime of brand A light bulbs has a width of less than 10 hours.

5 X is normally distributed with mean μ (unknown) and variance 100. A symmetric 95% confidence interval (145.2, 150.8) for μ was calculated from a random sample. Calculate the sample mean and the sample size.

7.2 Other confidence intervals

Confidence intervals for the mean of a normal distribution with unknown variance

In **7.1** it was established that, when $\bar{X}$ is normally distributed with mean μ and known variance σ^2, symmetric $100(1 - 2p)\%$ confidence limits for μ are given by

$$\bar{x} \pm z_p \text{SE}[\bar{X}]$$

where $\bar{x}$ is the mean of a random sample of size n and $\text{SE}[\bar{X}] = \dfrac{\sigma}{\sqrt{n}}$. Clearly, if σ^2 is unknown, $\text{SE}[\bar{X}]$ cannot be calculated and thus the confidence limits cannot be found. One possible solution is to replace $\text{SE}[\bar{X}]$ by an estimate calculated from the sample; an appropriate estimated standard error $\text{ESE}[\bar{X}]$ is $\dfrac{s}{\sqrt{n}}$, where s^2 is the sample unbiased estimate for σ^2. When n is large, s^2 is likely to be close to σ^2 and the limits

$$\bar{x} \pm z_p \text{ESE}[\bar{X}]$$

will have an associated confidence level close to $100(1 - 2p)\%$. However when n is small there is greater uncertainty and this case will be studied later in the chapter (see **7.6**).

Therefore it may be stated that, for large n (greater than 40 is a reasonable guideline), approximate $100\,(1 - 2p)\%$ confidence limits for μ are given by

$$\bar{x} \pm z_p \text{ESE}[\bar{X}]$$

where $\bar{x}$ is the sample mean and $\text{ESE}[\bar{X}] = \dfrac{s}{\sqrt{n}}$.

Example 3

The heights of men living in the U.K. are normally distributed with mean μ cm. The height x cm of each man in a random sample of 200 men living in the U.K. was measured and the following results were calculated:

$$\sum x = 35250, \qquad \sum x^2 = 6233409.$$

Calculate an approximate 95% confidence interval for μ.

$$\bar{x} = \frac{\sum x}{n}$$

$$\bar{x} = \frac{35250}{200} = 176.25$$

$$s^2 = \frac{\sum x^2 - n\bar{x}^2}{n - 1}$$

$$s^2 = \frac{6233409 - 200 \times 176.25^2}{199} = 103.5$$

$\bar{X}$ is normally distributed with mean μ and $\text{ESE}[\bar{X}] = \sqrt{\dfrac{103.5}{200}}$

Approximate 95% confidence limits for μ are given by

$$\bar{x} \pm 1.96 \times \text{ESE}[\bar{X}]$$

$$176.25 \pm 1.96 \times \sqrt{\frac{103.5}{200}}$$

$$176.25 \pm 1.41 \quad \text{(2 d.p.)}.$$

Approximate 95% confidence interval for μ is $(174.84, 177.66)$.

Confidence intervals for the difference of the means of two normal distributions

Let $\bar{X}_1$ be the mean of a random sample of n_1 observations from $N(\mu_1, \sigma_1{}^2)$ and let $\bar{X}_2$ be the mean of an independent random sample of n_2 observations from $N(\mu_2, \sigma_2{}^2)$.

The sampling distribution of $\bar{X}_1 - \bar{X}_2$ is normal with mean $\mu_1 - \mu_2$ and variance given by

$$V[\bar{X}_1 - \bar{X}_2] = V[\bar{X}_1] + V[\bar{X}_2] \qquad \bar{X}_1, \bar{X}_2 \text{ independent}$$

$$= \frac{\sigma_1^2}{n_1} + \frac{\sigma_2^2}{n_2}.$$

Thus, when the variances σ_1^2 and σ_2^2 are known, $100(1 - 2p)\%$ confidence limits for $(\mu_1 - \mu_2)$ are given by

$$(\bar{x}_1 - \bar{x}_2) \pm z_p SE[\bar{X}_1 - \bar{X}_2]$$

where $\bar{x}_1$ and $\bar{x}_2$ are the sample means and $SE[\bar{X}_1 - \bar{X}_2] = \sqrt{\frac{\sigma_1^2}{n_1} + \frac{\sigma_2^2}{n_2}}$.

Alternatively, when the variances σ_1^2 and σ_2^2 are unknown and n_1 and n_2 are large, approximate $100(1 - 2p)\%$ confidence limits are given by

$$(\bar{x}_1 - \bar{x}_2) \pm z_p ESE[\bar{X}_1 - \bar{X}_2]$$

where $ESE[\bar{X}_1 - \bar{X}_2] = \sqrt{\frac{s_1^2}{n_1} + \frac{s_2^2}{n_2}}$, and s_1^2 and s_2^2 are the sample unbiased estimates for σ_1^2 and σ_2^2, respectively.

Example 4

Repeated weighings of an object on a chemical balance give readings which are normally distributed with mean equal to the true weight of the object and standard deviation 0.5 mg. Twenty weighings of an object A gave a mean reading of 15.6 mg, while fifteen weighings of an object B gave a mean reading of 13.7 mg. Calculate a 90% confidence interval for the difference between the weights of A and B.

Let $\bar{X}_1$ be the mean of 20 readings of the weight of A, then $\bar{X}_1 \sim N\left(\mu_1, \frac{0.5^2}{20}\right)$.

Let $\bar{X}_2$ be the mean of 15 readings of the weight of B, then $\bar{X}_2 \sim N\left(\mu_2, \frac{0.5^2}{15}\right)$.

Therefore $SE[\bar{X}_1 - \bar{X}_2] = \sqrt{\left(\frac{0.5^2}{20} + \frac{0.5^2}{15}\right)} = 0.1707825$.

90% confidence limits for $\mu_1 - \mu_2$ are given by

$$(\bar{x}_1 - \bar{x}_2) \pm 1.645 \times SE(\bar{X}_1 - \bar{X}_2)$$

$$(15.6 - 13.7) \pm 1.645 \times 0.1707825$$

$$1.9 \pm 0.28 \quad (2 \text{ d.p.}).$$

90% confidence interval for the difference in the true weights of A and B is (1.62, 2.18).

Exercise 7.2

1 The masses of the contents of tins of rice pudding produced by a certain manufacturer are normally distributed. The mass, x g, of the contents of each of a random sample of 100 tins was measured and the following results calculated:

$$\sum x = 44196 \qquad \sum x^2 = 19533087.$$

Calculate unbiased estimates for the mean μ and the variance of the masses of the contents of tins of rice pudding produced by this manufacturer. Deduce an approximate 90% confidence interval for μ.

2 A random sample of 150 values from a normal population gave the following data:

$$\bar{x} = 81.4 \qquad \sum (x - \bar{x})^2 = 2628.36.$$

Calculate an unbiased estimate for the population variance and an approximate 98% confidence interval for the population mean.

3 A random sample of 9 observations from a normal population having a standard deviation of 4 had a mean of 122. A random sample of 16 observations from another normal population having a standard deviation of 5 had a mean of 101. Calculate a 99% confidence interval for the difference between the population means.

4 The lifetimes of bulbs of brand A and brand B are independent normally distributed random variables.

The lifetimes, measured in hours, of 100 bulbs of brand A had a sum of 98417 and the sum of their squares was 96948159. Calculate an approximate 90% confidence interval for the mean lifetime of bulbs of brand A.

The lifetimes, measured in hours, of 100 bulbs of brand B had a sum of 96324 and the sum of their squares was 92884506. Calculate an approximate 95% confidence interval for the difference in the mean lifetimes of bulbs of brand A and brand B.

5 The heights of men and women in a certain country are independent random variables.

The heights, measured in metres, of a random sample of 200 women had a sum of 316.00 and the sum of their squares was 499.9964. Calculate an approximate 95% confidence interval for the mean height of women in the country.

The heights, measured in metres, of a random sample of 150 men had a sum of 258.12 and the sum of their squares was 444.5454. Calculate an approximate 99% confidence interval for the difference between the mean heights of men and women in the country.

7.3 Approximate confidence intervals using the Central Limit Theorem

In the previous two sections it was assumed that the population distributions were normal, but, under some circumstances, this assumption is unnecessary. For example, suppose that the population distribution of X is unknown and that its mean is μ (unknown) and its variance σ^2 (known). Also suppose that $\bar{X}$ is the mean of a random sample of n observations of X. Provided that n is large, it follows from the Central Limit Theorem that the sampling distribution of $\bar{X}$ is approximately normal with mean μ and standard error $SE[\bar{X}] = \dfrac{\sigma}{\sqrt{n}}$. Using this approximation, it may be shown, as in **7.1**, that

$$\bar{x} \pm z_p SE[\bar{X}]$$

are approximate $100(1 - 2p)\%$ confidence limits for μ.

When σ is unknown, making a further approximation as in **7.2**,

$$\bar{x} \pm z_p ESE[\bar{X}]$$

where $ESE[\bar{X}] = \dfrac{s}{\sqrt{n}}$, are approximate $100(1 - 2p)\%$ confidence limits for μ.

The larger the value of n, the closer will be the actual confidence level of the intervals having the above limits to $100(1 - 2p)\%$. The approximation is usually quite good for values of n greater than 40, but it can be good for smaller values of n when the population distribution is symmetrical about its mean (or nearly so).

It also follows from the Central Limit Theorem that, provided n_1 and n_2 are large, the limits (using the notation defined on page 91)

$$(\bar{x}_1 - \bar{x}_2) \pm z_p ESE[\bar{X}_1 - \bar{X}_2]$$

will give an approximate $100(1 - 2p)\%$ confidence interval for $(\mu_1 - \mu_2)$, even when the population distributions are non-normal.

Example 5

A random sample of 200 men living in the U.K. gave unbiased estimates of 176 cm and 120 cm^2 for the mean μ_1 and the variance σ_1^2, respectively, of the heights of men living in the U.K. A random sample of 250 women living in the U.K. gave unbiased estimates of 162 cm and 90 cm^2 for the mean μ_2 and the variance σ_2^2, respectively, of the heights of women living in the U.K. Calculate approximate 90% confidence limits for $\mu_1 - \mu_2$.

Let $\bar{X}_1$ be the mean of a random sample of 200 men living in the U.K. and let $\bar{X}_2$ be the mean of a random sample of 250 women living in the U.K. Although nothing is known about either of the population distributions, the sampling distribution of $\bar{X}_1 - \bar{X}_2$ is approximately normal with mean $(\mu_1 - \mu_2)$ because the sample sizes are large.

$$\text{ESE}[\bar{X}_1 - \bar{X}_2] = \sqrt{\frac{s_1^2}{n_1} + \frac{s_2^2}{n_2}}$$

$$= \sqrt{\frac{120}{200} + \frac{90}{250}}$$

$$= 0.9797959$$

Approximate 90% confidence limits for $(\mu_1 - \mu_2)$ are given by

$$(\bar{x}_1 - \bar{x}_2) \pm 1.645 \times \text{ESE}[\bar{X}_1 - \bar{X}_2]$$

$$(176 - 162) \pm 1.645 \times 0.9797959$$

$$14 \pm 1.61 \quad (2\,\text{d.p.}).$$

An approximate 90% confidence interval for $(\mu_1 - \mu_2)$ is (12.39, 15.61).

Exercise 7.3

1 The yield, x kg, of each plant in a random sample of 100 tomato plants of a certain variety was measured and the following results calculated:

$$\sum x = 700 \text{ and } \sum x^2 = 5296.$$

Calculate unbiased estimates for the mean μ and the variance σ^2 of the yields from plants of this variety. Deduce an approximate 95% confidence interval for μ.

2 The breaking strength, x N, for each wire in a random sample of 200 was measured and the following results were calculated:

$$\sum x = 14240 \quad \text{and} \quad \sum x^2 = 1015679.$$

Calculate an approximate 90% confidence interval for the mean breaking strength of such wires.

3 A random sample of 500 was chosen from the workers living in a certain town in order to investigate the number, x, of days lost through illness. The following results were calculated:

$$\sum x = 5140 \qquad \sum x^2 = 102788.$$

Calculate an approximate 99% confidence interval for the mean number of days lost through illness by workers in the town.

4 For each packet in a random sample of 300 packets of tea packed by machine A the mass x g was measured and the following results calculated:

$$\sum x = 39600 \qquad \sum x^2 = 5231984.$$

Calculate unbiased estimates for the mean μ_A and the variance of the masses of all packets of tea packed by machine A. Determine an approximate 90% confidence interval for μ_A.

A random sample of 400 packets of tea packed by machine B gave unbiased estimates 135 g and 36 g^2 for the mean μ_B and the variance of the masses of packets of tea packed by machine B. Determine an approximate 90% confidence interval for $(\mu_A - \mu_B)$.

5 From a very large bundle of fibres a random sample of 200 fibres was selected. The breaking strength x of each fibre was measured and the following results were calculated:

$$\sum x = 3580.0 \qquad \sum x^2 = 64529.75.$$

Calculate unbiased estimates for the mean μ_1 and the variance of the breaking strengths of fibres in the bundle. Determine an approximate 95% confidence interval for μ_1.

After the remaining fibres in the bundle had been treated with a certain chemical, a second random sample of 200 fibres was taken. The unbiased estimates, calculated from this sample, for the mean μ_2 and the variance of the breaking strengths of the fibres in the treated bundle were 18.1 and 0.99, respectively. Determine an approximate 95% confidence interval for $(\mu_1 - \mu_2)$. Is there evidence, at this level of confidence, to suggest that the treatment has altered the mean breaking strength of fibres in the bundle?

7.4 Approximate confidence intervals for a population proportion

It was shown in 6.2 on page 75 that the sample proportion $\hat{P}$ was an unbiased estimator for the population proportion p and that its standard error was given by $SE[\hat{P}] = \sqrt{\dfrac{p(1-p)}{n}}$. Since $\hat{P}$ is the mean number of successes per trial for the n trials, it follows from the Central Limit Theorem that the distribution of $\hat{P}$ is approximately normal when n is large. Approximate $100(1-2p)\%$ confidence limits for the population proportion p are given by

$$\hat{p} \pm z_p SE[\hat{P}].$$

Since p is unknown, $SE[\hat{P}]$ cannot be evaluated, so it is usually replaced by an estimate, $ESE[\hat{P}] = \sqrt{\dfrac{\hat{p}(1-\hat{p})}{n}}$. Thus the limits

$$\hat{p} \pm z_p ESE[\hat{P}]$$

give an approximate $100(1-2p)\%$ confidence interval which is easy to evaluate. The approximation is good, provided that n is large and p is not close to 0 or 1.

Example 6

A manufacturer wishes to find a 95% confidence interval for the proportion p of defective items in a large batch produced by a certain machine.

a A random sample of 200 items are tested and 40 are found to be defective. Calculate an approximate 95% confidence interval for p.

b Find the smallest sample size which will ensure that the width of the confidence interval for p is less than 0.1.

a The sample proportion $\hat{p} = \dfrac{40}{200} = 0.2$

$$\text{ESE}[\hat{P}] = \sqrt{\frac{0.2(1 - 0.2)}{200}} = 0.0282842.$$

Approximate 95% confidence limits for p are given by

$$\hat{p} \pm 1.96 \times \text{ESE}[\hat{P}]$$

$$0.2 \pm 1.96 \times 0.0282842$$

$$0.2 \pm 0.055 \quad (3 \text{ d.p.}).$$

An approximate 95% confidence interval for p is $(0.145, 0.255)$

b The width w of the 95% confidence interval for p is given by

$$w = (\hat{p} + 1.96 \times \text{SE}[\hat{P}]) - (\hat{p} - 1.96 \times \text{SE}[\hat{P}])$$

$$w = 2 \times 1.96 \times \text{SE}[\hat{P}].$$

Therefore it is required that $2 \times 1.96 \times \text{SE}[\hat{P}] < 0.1$. In order to ensure this holds whatever the value of p, it is necessary to replace $\text{SE}[\hat{P}]$ by its upper limit $\sqrt{\left(\dfrac{1}{4n}\right)}$, see page 75.

$$2 \times 1.96 \times \sqrt{\frac{1}{4n}} < 0.1$$

$$2 \times \frac{1.96}{0.1} \times \sqrt{\frac{1}{4}} < \sqrt{n}$$

$$19.6 < \sqrt{n}$$

$$384.16 < n$$

Therefore the required value of n is 385.

If, as in this question, there is reason to believe that p is near to 0.2, then a less conservative value for n may be calculated by replacing $\text{SE}[\hat{P}]$ by its estimated value $\text{ESE}[\hat{P}]$ instead of by its upper limit. However this value for n will not *guarantee* that the width of the confidence interval will be less than 0.1.

More accurate method for Example 6a

In part **a** above, the method used two approximations, viz. that the distribution of $\hat{P}$ is approximately normal and that $ESE[\hat{P}] \simeq SE[\hat{P}]$. The following method avoids the use of the second of these approximations.

Since the distribution of $Z = \dfrac{\hat{P} - p}{SE[\hat{P}]}$ is approximately $N(0, 1)$

$$P\left(-1.96 < \frac{\hat{P} - p}{\sqrt{\dfrac{p(1 - p)}{n}}} < 1.96 \right) = 0.95$$

and the 95% confidence limits for p will be given by

$$\frac{\hat{p} - p}{\sqrt{\dfrac{p(1 - p)}{n}}} = \pm 1.96.$$

In this question $\hat{p} = 0.2$ and $n = 200$

$$\frac{0.2 - p}{\sqrt{\dfrac{p(1 - p)}{200}}} = \pm 1.96.$$

Clearing fractions $\qquad\qquad (0.2 - p)\sqrt{200} = +1.96\sqrt{p(1 - p)}.$

Squaring $\qquad\qquad\qquad\quad (0.2 - p)^2\, 200 = 3.8416 p(1 - p).$

Clearing brackets $\qquad 200(0.04 - 0.4p + p^2) = 3.8416p - 3.8416p^2$

$$8 - 80p + 200p^2 = 3.8416p - 3.8416p^2$$

$$203.8416p^2 - 83.8416p + 8 = 0$$

$$p = \frac{83.8416 \pm \sqrt{83.8416^2 - 4 \times 203.8416 \times 8}}{2 \times 203.8416}$$

$$p = 0.150 \quad \text{or} \quad 0.261 \quad (3\,\text{d.p.}).$$

An approximate 95% confidence interval for p is $(0.150, 0.261)$

This interval, which is not symmetric, has a confidence level closer to 95% than the one, $(0.145, 0.255)$, calculated earlier, but it is much more difficult to evaluate.

Example 7

A random sample of 846 fish was taken from a lake, marked, and returned to the lake. Some time later a second random sample of 300 fish was taken and of these 36 were found to be marked.

a Estimate the number N of the fish in the lake.

b Obtain approximate 90% confidence limits for the proportion p of marked fish in the lake.

c Deduce approximate 90% confidence limits for N, giving your answers to the nearest hundred.

a Estimate for p: $\hat{p} = \dfrac{36}{300} = 0.12$.

Estimate for N: $\hat{N} = \dfrac{846}{0.12} = 7050$.

b

$$\text{ESE}[\hat{P}] = \sqrt{\dfrac{\hat{p}(1-\hat{p})}{n}}$$

$$\text{ESE}[\hat{P}] = \sqrt{\dfrac{0.12 \times (1-0.12)}{300}}$$

$$= 0.0187616$$

Approximate 90% confidence limits for p are given by

$$\hat{p} \pm 1.645 \times \text{ESE}[\hat{P}]$$

$$0.12 \pm 1.645 \times 0.0187616$$

$$0.12 \pm 0.031 \quad (3\,\text{d.p.}).$$

An approximate 90% confidence interval for p is (0.089, 0.151).

c Approximate 90% confidence limits for N are obtained by dividing 846 by the corresponding limits for p.

$$\text{Upper limit} = \dfrac{846}{0.089} = 9500 \text{ (to nearest 100)}$$

$$\text{Lower limit} = \dfrac{846}{0.151} = 5600 \text{ (to nearest 100)}$$

The approximate 90% confidence interval for N, (5600, 9500), is not symmetric as the point estimate 7050 is not the mid-point of the interval.

Approximate confidence intervals of the difference between two population proportions

Example 8

In an opinion poll 540 out of a total of 1200 interviewed said that they supported the government. One month later a second opinion poll, independent of the first, revealed that 390 out of 1000 supported the government. Calculate an approximate 99% confidence interval for $p_1 - p_2$, where p_1, p_2 denote respectively the proportions of the electorate that support the government at the time of the first and second polls.

$$\hat{p}_1 = \dfrac{540}{1200} = 0.45$$

$$\hat{p}_2 = \frac{390}{1000} = 0.39$$

$\hat{P}_1 - \hat{P}_2$ is approximately normally distributed with mean $(p_1 - p_2)$ and variance given by

$$V[\hat{P}_1 - \hat{P}_2] = V[\hat{P}_1] + V[\hat{P}_2] \qquad \hat{P}_1, \hat{P}_2 \text{ independent}$$

$$= \frac{p_1(1 - p_1)}{1200} + \frac{p_2(1 - p_2)}{1000}$$

$$\mathrm{ESE}[\hat{P}_1 - \hat{P}_2] = \sqrt{\frac{0.45 \times (1 - 0.45)}{1200} + \frac{0.39 \times (1 - 0.39)}{1000}}$$

$$= 0.0210748.$$

Approximate 99% confidence limits for $(p_1 - p_2)$ are given by

$$(\hat{p}_1 - \hat{p}_2) \pm 2.576 \times \mathrm{ESE}[\hat{P}_1 - \hat{P}_2]$$

$$(0.45 - 0.39) \pm 2.576 \times 0.0210478$$

$$0.06 \pm 0.054 \quad (3 \text{ d.p.}).$$

An approximate 99% confidence interval for $(p_1 - p_2)$ is $(0.006, 0.114)$.

Since this interval does not contain 0, it suggests that there has been a change in the government's level of support. Furthermore since each limit is positive, the interval suggests that there has been a decrease in the level of support for the government.

Exercise 7.4

1 From a random sample of 1000 houses in a large town it was found that 427 had a video recorder. Calculate an approximate 95% confidence interval for the proportion of all houses in the town that have a video recorder.

2 In a random sample of 200 voters in a certain constituency 84 of them said that they would vote for party A in a forthcoming election.

   a Calculate an approximate 90% confidence interval for the proportion of the electorate in the town that will vote for party A in the election.

   b Estimate the minimum sample size required to obtain a 90% confidence interval having a width of 0.02.

   c Find the smallest sample size which will ensure that the width of a 90% confidence interval is less than 0.02.

3 From the 2500 pupils at a school a random sample of 50 was chosen and it was found that 7 of these pupils were left-handed. Find approximate 90% confidence intervals for

   a the proportion of pupils in the school who are left-handed

   b the number of the pupils in the school who are left-handed.

4   A random sample of 400 rabbits in a particular area was caught, tagged and released in the same area. Later, a second random sample of 360 rabbits was caught and of these 18 were found to be tagged.

   a   Estimate the number N of rabbits in the area.

   b   Calculate approximate 95% confidence limits for the proportion of tagged rabbits in the area.

   c   Find approximate 95% confidence limits for N.

5   In a random sample of 1000 persons available for work there were 620 men and 380 women. Of the men 93 were unemployed, while 46 of the women were unemployed. Calculate approximate 99% confidence limits for

   a   the proportion of unemployed men

   b   the proportion of unemployed women

   c   the difference between the proportions of unemployed men and women.

7.5 Approximate confidence intervals for the mean of a Poisson distribution

Suppose that $\bar{X}$ is the mean of n observations from a Poisson distribution with unknown mean α. Since the variance of a Poisson distribution is equal to its mean, it follows from the Central Limit Theorem that, provided n is large, the sampling distribution of $\bar{X}$ is approximately normal with mean α and standard error $\text{SE}[\bar{X}] = \sqrt{\left(\dfrac{\alpha}{n}\right)}$. As before, since $\text{SE}[\bar{X}]$ cannot be evaluated, approximate $100(1 - 2p)\%$ confidence limits for α are given by

$$\bar{x} \pm z_p \text{ESE}[\bar{X}].$$

In this case, since $\bar{x}$ is an unbiased estimate for α, $\text{ESE}[\bar{X}] = \sqrt{\left(\dfrac{\bar{x}}{n}\right)}$ is an appropriate estimate for the standard error of $\bar{X}$.

Example 9

The number of faults per car made on a particular production line is known to have a Poisson distribution with an unknown mean α. An inspection of 100 of these cars revealed a total of 810 faults. Calculate an approximate 99% confidence interval for α.

$$\bar{x} = \frac{810}{100} = 8.1$$

$$\text{ESE}[\bar{X}] = \sqrt{\frac{\bar{x}}{n}}$$

$$= \sqrt{\frac{8.1}{100}}$$

$$= 0.2846049$$

Approximate 99% confidence limits for α are given by

$$\bar{x} \pm 2.576 \times \text{ESE}[\bar{X}]$$

$$8.1 \pm 2.576 \times 0.2846049$$

$$8.1 \pm 0.73 \quad (2\,\text{d.p.})$$

An approximate 99% confidence interval for α is (7.37, 8.83).

More accurate method for Example 9

Since the distribution of $Z = \dfrac{\bar{X} - \alpha}{\text{SE}[\bar{X}]}$ is approximately $N(0, 1)$

$$P\left(-2.576 < \frac{\bar{X} - \alpha}{\sqrt{\left(\dfrac{\alpha}{n}\right)}} < 2.576 \right) = 0.99$$

and the 99% confidence limits for α are given by

$$\frac{\bar{x} - \alpha}{\sqrt{\left(\dfrac{\alpha}{n}\right)}} = \pm 2.576$$

In this question $\bar{x} = 8.1$ and $n = 100$

$$\frac{8.1 - \alpha}{\sqrt{\left(\dfrac{\alpha}{100}\right)}} = \pm 2.576$$

Clearing fractions $\qquad\qquad (8.1 - \alpha)\sqrt{100} = \pm 2.576\sqrt{\alpha}$

Squaring $\qquad\qquad\qquad (8.1 - \alpha)^2 100 = 6.635776\alpha$

Clearing brackets $\qquad (65.61 - 16.2\alpha + \alpha^2)100 = 6.635776\alpha$

$$6561 - 1620\alpha + 100\alpha^2 = 6.635776\alpha$$

$$100\alpha^2 - 1626.635776\alpha + 6561 = 0$$

$$\alpha = \frac{1626.635776 \pm \sqrt{1626.635776^2 - 4 \times 100 \times 6561}}{2 \times 100}$$

$$\alpha = 7.40 \text{ or } 8.87 \quad (2 \text{ d.p.})$$

An approximate 99% confidence interval for α is (7.40, 8.87).

The interval, which is not symmetric, has a confidence level closer to 99% than the one calculated earlier, but it is more difficult to evaluate.

Exercise 7.5

1 The number of particles emitted per minute by a radioactive source has a Poisson distribution with unknown mean α. The total number of particles emitted during 100 randomly selected one-minute intervals was 754. Find an approximate 90% confidence interval for α.

2 The number of breakdowns per day in summer on a certain stretch of motorway may be assumed to have a Poisson distribution with unknown mean α. The total number of breakdowns in June was 123. Calculate approximate 95% confidence limits for α, stating any further assumption made.

3 The number of claims per week made on life insurance policies of a certain company has a Poisson distribution with mean μ. If 728 claims were made last year, calculate approximate 95% confidence limits for μ.

4 Re-calculate the confidence limits required in question **3** using the alternative method described on page 101.

5 A total of 348 flaws were found in a random sample of 200 rolls of material produced by machine A and a total of 236 flaws were found in a random sample of 200 rolls produced by machine B. Determine an approximate 99% confidence interval for the difference between the mean numbers of flaws per roll produced by machines A and B.

7.6 Student's *t*-distribution

If $\bar{X}$ is the mean of a random sample of n observations from a $N(\mu, \sigma^2)$ distribution then the sampling distribution of $\bar{X}$ is $N\left(\mu, \dfrac{\sigma^2}{n}\right)$. It follows that the random variable

$$Z = \frac{\bar{X} - \mu}{\dfrac{\sigma}{\sqrt{n}}}$$

has a $N(0, 1)$ distribution. This result was used in **7.1** to obtain confidence limits for μ when σ was known. When the value of σ is unknown, as is often the case, it is usually replaced by the value of S, where S^2 is the sample unbiased estimator for the variance σ^2. In **7.2**, when the sample size n was large, the random variable

$$T = \frac{\bar{X} - \mu}{\dfrac{S}{\sqrt{n}}}$$

was considered to have an approximate $N(0, 1)$ distribution in order to obtain approximate confidence limits for μ. However the distribution of T is not normal and, when n is small, the normal approximation is not sufficiently accurate. The probability density function of T was discovered by the statistician W. S. Gossett, who published his results under the pseudonym 'Student', presumably to hide the fact that he was working for the brewery Guinness at the time.

The probability density function f of T is given by

$$f(t) = k\left(1 + \frac{t^2}{n-1}\right)^{-\frac{1}{2}n}, \qquad -\infty < t < \infty,$$

where k is a constant which ensures that the area under the graph of f is unity. T is said to have a t-distribution with $(n-1)$ degrees of freedom.

For each value of n, the graph of f is continuous, symmetrical and similar in shape to the graph of ϕ, the probability density function of the standard normal variable; the graph of f is less peaked and has longer tails than the graph of ϕ. The diagram below shows the graphs of ϕ and two t-distributions (for $n = 5$ and $n = 10$). As n increases the graph of f approaches that of ϕ.

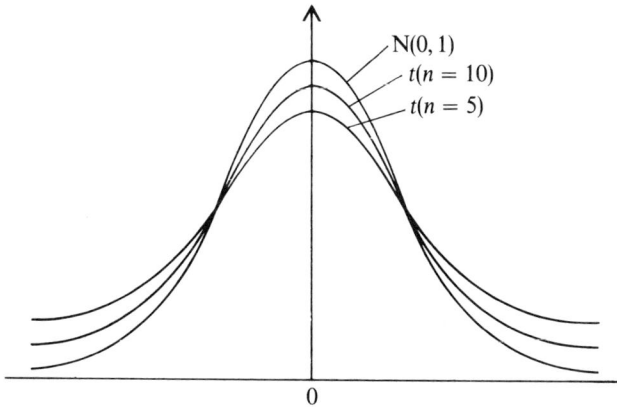

Confidence intervals for the mean of a normal distribution with unknown variance

To obtain $100(1 - 2p)\%$ confidence intervals for μ when σ is unknown, a method similar to that used in **7.1** is adopted but using the distribution of T instead of Z.

Let t_p denote the value such that
$$P(T > t_p) = p$$

By symmetry $$P(T < -t_p) = p$$

Combining $$P(-t_p < T < t_p) = 1 - 2p$$

Therefore $$P\left(-t_p < \frac{\bar{X} - \mu}{\frac{S}{\sqrt{n}}} < t_p\right) = 1 - 2p$$

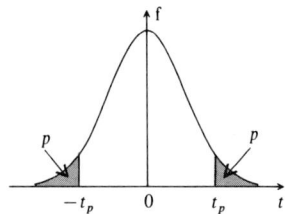

Re-arranging
$$P\left(\bar{X} - t_p\frac{S}{\sqrt{n}} < \mu < \bar{X} + t_p\frac{S}{\sqrt{n}}\right) = 1 - 2p$$

Therefore, when $\bar{x}$ and s denote the observed values of $\bar{X}$ and S, the symmetrical $100(1 - 2p)\%$ confidence limits are given by

$$\bar{x} \pm t_p\frac{s}{\sqrt{n}}.$$

In order to evaluate these limits it is necessary to know the values for t_p. **Table 4**, (page 232) lists values of t_p for some values of p and v, where the Greek letter v (nu) represents the number of degrees of freedom. For the confidence limits under consideration $v = n - 1$. **Table 4** was compiled by numerical methods and it may be noted that the entries in the last row (with $v = \infty$) are identical to the values of z_p used previously.

Example 10

Ten university students independently conducted experiments to determine the value of g, where $g\,ms^{-2}$ is the acceleration due to gravity. They obtained the following results:

9.812,	9.807,	9.804,	9.805,	9.812,
9.808,	9.807,	9.814,	9.809,	9.807.

Calculate 95% confidence limits for g, stating any assumptions that you make.

Using a calculator $\qquad \sum x = 98.085 \qquad\qquad \sum x^2 = 962.066817$

$$\bar{x} = \frac{\sum x}{n}$$

$$\bar{x} = 98.085/10 = 9.8085$$

$$s^2 = \frac{\sum x^2 - n\bar{x}^2}{n - 1}$$

$$s^2 = \frac{962.066817 - 10 \times 9.8085^2}{9} = 1.05 \times 10^{-5}$$

$$s = 0.00324 \quad (3\text{ s.f.})$$

The number of degrees of freedom $v = n - 1 = 9$ and for 95% confidence limits $p = 0.025$. The required value of $t_p = 2.26$ is found in the 9th row and the 2nd column of **Table 4**.

The 95% confidence limits for g are given by

$$\bar{x} \pm t_p\frac{s}{\sqrt{n}}$$

$$9.8085 \pm 2.26 \times \frac{0.00324}{\sqrt{10}}$$

$$9.8085 \pm 0.0023(4\,\text{d.p.})$$

The symmetrical 95% confidence interval for g is (9.8062, 9.8108).

It has been assumed that the experimental results are normally distributed with mean g and that the 10 results obtained form a random sample.

Exercise 7.6

1 The volume of wine poured into bottles by an automatic filling machine may be assumed to be normally distributed. An inspector obtains a random sample of 10 bottles and measures the volume x cl of wine in each bottle. The following results were calculated:

$$\sum x = 710, \qquad \sum x^2 = 50428.$$

Find a 90% confidence interval for the mean volume of wine delivered by the machine.

2 The heights (measured in cm) of 15 guardsmen are

183,   192,   190,   184,   187,   191,   195,   186,
188,   190,   192,   190,   184,   185,   192.

Calculate a 95% confidence interval for the mean height of guardsmen, stating any assumption you have made.

3 In an experiment to test the weight-reducing effect of a diet, eight women were weighed before and after following the diet for a month. Their weight losses, in kg, were

2.4,   3.5,   3.4,   1.1,   3.3,   4.6,   5.2,   1.3.

Find a symmetrical 99% confidence interval for the mean loss in weight for women following the diet for a month. State, with a reason, whether or not your interval supports the claim that the average weight loss for women following the diet for a month is 3.5 kg.

4 In order to estimate the mean lifetime of an electronic component, a manufacturer carried out tests on a random sample of 21 of these components. From these tests the following results were obtained:

$$\sum x = 945, \qquad \sum x^2 = 42605,$$

where x hours is the lifetime of a sampled component. Assuming that the lifetimes are normally distributed, find a 98% confidence interval for the mean lifetime of the electronic components made by the manufacturer.

5 From a random sample of 9 observations from a $N(\mu, \sigma^2)$ distribution, a symmetric 90% confidence interval for μ was calculated to be (10.76, 13.24). Find unbiased estimates for μ and σ^2.

7.7 Further examples

Example 11

A continuous random variable X is uniformly distributed between 5 and θ, where $\theta > 5$, but is otherwise unknown. Write down, in terms of θ, the mean μ of X.

From a random sample of 100 observations of X the following results were calculated:

$$\sum x = 812.0, \qquad \sum x^2 = 6816.19.$$

a Find an approximate 95% confidence interval for μ.

b Deduce an approximate 95% confidence interval for θ.

When $X \sim \mathrm{U}(a, b)$ $\qquad E[X] = \dfrac{(a + b)}{2}$

Since $X \sim \mathrm{U}(5, \theta)$ $\qquad \mu = \dfrac{(5 + \theta)}{2}$

a $\bar{x} = \dfrac{\sum x}{n}$

$x = \dfrac{812.0}{100} = 8.12$

$s^2 = \dfrac{\sum x^2 - n\bar{x}^2}{n - 1}$

$s^2 = \dfrac{6816.19 - 100 \times 8.12^2}{99} = 2.25$

$s = 1.5$

Since n is large, the distribution of $\bar{X}$ is approximately $\mathrm{N}\left(\mu, \dfrac{\sigma^2}{n}\right)$ (the Central Limit Theorem). Therefore approximate 95% confidence limits for μ are given by

$$\bar{x} \pm 1.96 \times \mathrm{ESE}[\bar{X}]$$

$$8.12 \pm 1.96 \times \dfrac{1.5}{\sqrt{100}}$$

$$8.12 \pm 0.294 \quad (3\,\text{d.p.})$$

An approximate 95% confidence interval for μ is (7.826, 8.414)

b $\mu = \dfrac{(5 + \theta)}{2}$ $\qquad \theta = 2\mu - 5$

Approximate 95% confidence limits for θ are given by

Upper limit $= 2 \times 8.414 - 5 = 11.828$

Lower limit $= 2 \times 7.826 - 5 = 10.652$

An approximate 95% confidence interval for θ is (10.652, 11.828)

Example 12

$\bar{X}$ is the mean of a random sample of 64 observations from a normal distribution whose mean is λ and whose standard deviation is also λ, where λ is a positive constant.

a Find, in terms of λ, the values of a and b such that
$$P(\bar{X} < a) = P(\bar{X} > b) = 0.025.$$

b Given that $\bar{x} = 29$ deduce 95% confidence limits for λ.

a $X \sim N(\lambda, \lambda^2)$ $\qquad\qquad \bar{X} \sim N\left(\lambda, \dfrac{\lambda^2}{64}\right)$

Let $Z = \dfrac{X - \lambda}{\dfrac{\lambda}{8}} \sim N(0, 1)$

$P(\bar{X} < a) = 0.025$ $\qquad\qquad\qquad$ $P(\bar{X} > b) = 0.025$

$P\left(Z < \dfrac{a - \lambda}{\dfrac{\lambda}{8}}\right) = 0.025$ $\qquad\qquad$ $P\left(Z > \dfrac{b - \lambda}{\dfrac{\lambda}{8}}\right) = 0.025$

$\dfrac{a - \lambda}{\dfrac{\lambda}{8}} = -1.96$ $\qquad\qquad\qquad$ $\dfrac{b - \lambda}{\dfrac{\lambda}{8}} = 1.96$

$a - \lambda = -1.96 \times \dfrac{\lambda}{8}$ $\qquad\qquad$ $b - \lambda = 1.96 \times \dfrac{\lambda}{8}$

$a - \lambda = -0.245\lambda$ $\qquad\qquad\qquad$ $b - \lambda = 0.245\lambda$

$a = 0.755\lambda$ $\qquad\qquad\qquad\qquad$ $b = 1.245\lambda$

b $\qquad\qquad P(a < \bar{X} < b) = 0.95$

$P(0.755\lambda < \bar{X} < 1.245\lambda) = 0.95$

$P\left(\dfrac{\bar{X}}{1.245} < \lambda < \dfrac{\bar{X}}{0.755}\right) = 0.95$

When $\bar{x} = 29$, the upper 95% confidence limit for $\lambda = \dfrac{29}{0.755} = 38.4$

the lower 95% confidence limit for $\lambda = \dfrac{29}{1.245} = 23.3$

A 95% confidence interval for λ is (23.3, 38.4).

Miscellaneous Exercise 7

1   In an opinion poll conducted in a certain constituency it was found that 468 of a random sample of 900 voters stated independently that they would be voting for party A in a forthcoming by-election. Calculate an approximate 95 per cent symmetric confidence interval for the proportion of the electorate in the constituency that will vote for party A in the by-election. State one assumption that has to be made about the responses obtained in the poll for your calculation to be valid.   (*JMB*)

2   From a lake, a random sample of 500 fish is caught, marked and returned to the lake. Later, in a random sample of 400 fish, taken with replacement, 80 are found to be marked. Estimate the number of fish in the lake. Find an approximate 90 per cent symmetric confidence interval for the proportion of marked fish in the lake. Deduce approximate 90 per cent confidence limits for the number of fish in the lake, giving your answers to the nearest hundred.   (*JMB*)

3   An experimenter wishes to determine the value of a physical constant. From experience with the apparatus used, he knows that the observations obtained will be normally distributed, with mean μ equal to the true value of the constant and with variance 0.0075. He takes 27 observations using the apparatus and calculates their mean to be 3.874. Determine 95 per cent confidence limits for the true value of the constant.   (*JMB*)

4   The yields, x kg, of tomatoes from a random sample of 40 plants of a particular variety gave the values

$$\sum x_i = 240, \qquad \sum x_i^2 = 1596.$$

Calculate unbiased estimates of the mean μ and the variance σ^2 of the yields from plants of this variety.

Assuming that your estimate of σ^2 is the true value, determine an approximate 95 per cent confidence interval for μ.   (*JMB*)

5   In a random sample of 200 resistors of a particular brand it was found that 13 of them failed to meet the specified tolerance. Denoting by p the proportion of all such resistors that fail to meet the specified tolerance, write down an unbiased estimate of p and obtain an estimate of its standard error. Hence calculate an approximate 95% confidence interval for the value of p. State whether or not your result supports the manufacturer's claim that $p = 0.04$.

Assuming that the manufacturer's claim ($p = 0.04$) is correct, use a distributional approximation to calculate the probability that in a random sample of 200 resistors at least 13 will fail to meet the specified tolerance.   (*JMB*)

6 A company produces electrical elements.

    a A random sample of 80 of these elements had resistances $x_1, x_2, \ldots, x_{80}$ ohms, where

$$\sum x_i = 790 \qquad \text{and} \qquad \sum x_i^2 = 7821.$$

       (i) Calculate unbiased estimates of the mean and the variance of the resistances of elements produced by the company.

      (ii) Use a normal approximation to calculate, correct to three decimal places, 95% symmetrical confidence limits for the mean resistance of elements produced by the company.

    b Suppose that the standard deviation of the resistances of elements produced by the company is equal to 0.5 ohms. Determine the smallest number of elements that should be sampled for there to be a probability of at least 0.95 that the *sample* mean resistance will be within 0.05 ohms of the mean resistance of elements produced by the company. *(JMB)*

7 In a random sample of 100 observations of a continuous random variable X it was found that 64 of them had values greater than 3. Use this information to obtain an unbiased estimate of θ, the proportion of the values of X that are greater than 3, and an estimate of its standard error. Hence find approximate 95% confidence limits for θ, giving each limit to two decimal places.

Suppose that X is uniformly distributed over the interval from $\alpha - 1$ to $\alpha + 1$, where α is an unknown constant. Obtain an expression for θ in terms of α. Hence find approximate 95% confidence limits for α giving each limit to two decimal places. *(JMB)*

8 The random variable X is normally distributed with unknown mean μ cm and standard deviation 2 cm.

    (i) A random sample of 16 observations of X had values which summed to 118.4 cm. Calculate a 95% confidence interval for the value of μ.

    (ii) Find the smallest sample size, n, of observations of X that should be taken for the width of the 95% confidence interval for μ calculated from the sample values to be less than 1 cm. *(WJEC)*

9 The continuous random variable X has probability density function f, where

$$f(x) = 1, \qquad \theta + 1 \leqslant x \leqslant \theta + 2,$$
$$f(x) = 0, \qquad \text{otherwise,}$$

where θ is an unknown constant. Denoting the mean and the variance of X by μ and σ^2, respectively, express μ in terms of θ, and find the value of σ^2.

A random sample of 10 observations of X had the values:

      2.4,   1.6,   1.7,   2.3,   1.9,   1.8,   1.6,   2.1,   1.6,   2.0.

Calculate an unbiased estimate of μ, and deduce an unbiased estimate of θ. Calculate, to three significant figures, the standard error of your unbiased estimate of θ.

Given that a random sample of 100 observations of X had mean 1.86, use a normal distribution approximation to calculate 95% confidence limits for θ, giving each limit correct to three significant figures.   (WJEC)

10 If $\bar{X}$ is the mean of a random sample of size n_1 from a population distribution having mean μ_1 and variance $\sigma_1{}^2$, and n_1 is very large, what can you say about the sampling distribution of $\bar{X}$?

If $\bar{Y}$ is the mean of a random sample of size n_2 from another population distribution having mean μ_2 and variance $\sigma_2{}^2$, and n_2 is also very large, what can you say about the sampling distribution of $\bar{Y} - \bar{X}$?

The heights, measured in metres, of 100 women drawn at random from a certain tribe had a sum of 160 and the sum of their squares was 265. Use this information to construct a 90 per cent confidence interval for the mean height of women of this tribe.

The heights, measured in metres, of 200 men drawn at random from the same tribe, had a sum of 360 and the sum of their squares was 712. Use the information contained in the two samples to construct a 95 per cent confidence interval for the amount by which the average height of the men in the tribe exceeds that of the women.   (WJEC)

11 Each of five pupils performed 10 independent trials of an experiment to determine the value of a physical constant. The means of the 10 observed values obtained by the five pupils were, respectively,

$$28.72, \qquad 29.01, \qquad 28.48, \qquad 28.63, \qquad 28.76.$$

It may be assumed that experimentally determined values of the physical constant are independent and normally distributed with mean equal to the true value of the constant and with variance σ^2. Use all five means given above to obtain

(i)   an unbiased estimate of the true value of the physical constant,

(ii) an unbiased estimate of the value of σ^2.

Assuming that $\sigma^2 = 0.36$, calculate 95% confidence limits for the true value of the physical constant.   (JMB)

12 A random sample of 16 observations is to be drawn from a normal distribution whose mean and standard deviation are both equal to λ, where λ is an unknown positive constant. Denoting the sample mean by $\bar{X}$, find the values of a and b, correct to two decimal places, such that

$$P(\bar{X} < a\lambda) = P(\bar{X} > b\lambda) = 0.025.$$

Given that the observed sample mean is 2.6, deduce 95 per cent confidence limits for the value of λ.   (JMB)

13 Let $\bar{X}$ denote the mean of a random sample of n observations from a normal distribution whose mean is λ and whose standard deviation is also λ, where λ is an unknown positive constant.
(i) Find the smallest value of n for which there is a probability of at least 0.95 that the numerical difference between $\bar{X}$ and λ will be less than 0.1λ.
(ii) For the case when $n = 100$, find a and b, in terms of λ, such that
$$P(\bar{X} < a) = P(\bar{X} > b) = 0.025.$$

Given that in this case $\bar{X} = 55$, deduce a 95% confidence interval for λ, giving each limit of the interval correct to the nearest integer.

(JMB)

14 A process extracts two chemicals A and B from 1 kg batches of raw material. The amount of A extracted from a batch is X grams, where X is a continuous random variable having mean μ and standard deviation 2, and the amount of B extracted is Y grams, where Y is a continuous random variable, independent of X, having mean λ and standard deviation 3. Chemical A is valued at £2 per gram, while chemical B is valued at £1 per gram. The process was applied to 100 batches, the means of the amounts of A and B extracted per batch being 5.2 grams and 8.6 grams, respectively.

Calculate approximate 95% confidence limits for the true mean combined value of the chemicals extracted per batch.

(JMB)

15 A continuous random variable X is known to have a uniform distribution in the interval $(10, a)$, where $a > 10$, but is otherwise unknown. Write down an expression for f(x), where f is the probability density function of X. Write down, in terms of a, the mean of X. Show that the variance of X is $\dfrac{(a - 10)^2}{12}$.

In one observation taken at random from this distribution a value of 21 for X was obtained. Write down an unbiased estimate of a. In 50 random observations of X the following results were obtained:
$$\sum x_i = 1100, \qquad \sum x_i^2 = 35000.$$

Use the value of $\sum x_i$ to obtain an unbiased estimate of a and give an approximate 95% symmetric confidence interval for a.

Given that one of the 50 observations of X had value 60, state why a is certain to be greater than or equal to 60. Since 60 is outside the calculated confidence interval explain the apparent contradiction between this certainty and the 95% confidence interval.

(JMB)

16 A random sample of eight observations from a normal distribution gave the values
$$1.6, \quad 1.3, \quad 2.1, \quad 1.8, \quad 1.6, \quad 1.1, \quad 1.9, \quad 2.2.$$
(i) Calculate unbiased estimates of the mean and the variance of the distribution.
(ii) Determine a 95% confidence interval for the mean of the distribution.

(WJEC)

17 The random variable X is normally distributed with unknown mean μ cm and unknown standard deviation σ cm.

  (i) A random sample of six observations of X had the values (in cm)

$$13.2, \quad 16.6, \quad 15.7, \quad 16.1, \quad 14.3, \quad 15.9,$$

respectively. Calculate 90% confidence limits for μ.

  (ii) Assuming that $\sigma = 1.6$ calculate, to four decimal places, the probability that the mean of a random sample of 16 observations of X will be within 1 cm of the value of μ. *(WJEC)*

18 a When an object is weighed on a chemical balance the readings obtained are subject to random errors which are known to be independent and normally distributed with mean zero and standard deviation 1 mg. A certain object is to be weighed 9 times on such a balance and the mean of the 9 readings is to be calculated. Find the probability that the mean of the 9 readings will be within 0.5 mg of the true weight of the object.

b Another weighing device is undergoing tests to determine its accuracy. A certain object of known true weight 50 mg was weighed 10 times on this device and the readings in mg were

$$49, \quad 51, \quad 49, \quad 52, \quad 49, \quad 50, \quad 52, \quad 51, \quad 49, \quad 48.$$

  (i) Calculate an unbiased estimate of the variance of the errors in readings using this device.

  (ii) Calculate 95 per cent confidence limits for the mean error in readings using this device. *(WJEC)*

19 a In an experiment to test the weight-reducing effect of a new diet, 6 women were weighed before and after following the diet. Their weight losses, in kg, were

$$1.6, \quad 0.8, \quad 0.1, \quad 1.4, \quad 0.5, \quad 0.7.$$

Stating clearly any assumptions that you make, calculate a 95% confidence interval for the mean loss in weight for women following the diet. State whether or not your interval supports the claim that the average loss in weight for women following this diet is 1 kg.

b In a large-scale experiment for comparing two diets A and B, 100 women followed diet A and 80 women followed diet B. From the weight losses of the 100 women who followed diet A it was concluded that the unbiased estimate of the mean weight loss was 2.97 kg and the unbiased estimate of the variance of the weight losses was 1.62 kg^2. From the weight losses of the 80 women who followed diet B the corresponding unbiased estimates were 2.31 kg and 1.55 kg^2, respectively. Stating clearly any assumptions that you make, calculate an approximate 95% confidence interval for the difference between the mean losses in weight for the two diets. What can you say about the relative weight-reducing effects of the two diets? *(WJEC)*

20 a The drained weights, in grammes, of a random sample of 8 cans of fruit of a particular brand were found to be

$$342, \quad 344, \quad 340, \quad 339, \quad 341, \quad 338, \quad 341, \quad \text{and} \quad 343,$$

respectively. Given that the drained weights are normally distributed, calculate a 90% confidence interval for the mean drained weight of fruit per can of this particular brand. State, with your reason, whether or not your result is consistent with the claim on the can that the average weight of fruit is 340 g.

b When an object is weighed on a certain weighing scale its recorded weight, in grammes, is a random value from a normal distribution whose mean is the true weight of the object and whose standard deviation is 0.2 g.

(i) Given that the mean of nine independently recorded weights of a particular object was 7.1 g, calculate a 99% confidence interval for the true weight of the object.

A second object was weighed sixteen times on the same scale and the mean of the recorded weights was found to be 8.4 g.

(ii) Determine a 95% confidence interval for the difference between the true weights of the two objects. *(WJEC)*

21 a Five independent measurements of the diameter of a ball bearing were made using a certain instrument. The results obtained, in millimetres, were:

$$8.9, \quad 9.1, \quad 9.1, \quad 8.9, \quad 8.9.$$

(i) Given that the true diameter of the ball bearing is 9 mm, calculate unbiased estimates of the mean and the variance of the measurement error of the instrument.

(ii) Assuming that the measurement errors are independent and normally distributed, calculate 90% confidence limits for the mean measurement error.

b The weekly wages received by a random sample of 80 personnel employed in factory *A* had a mean of £86·40 and a standard deviation of £3·80. Use this information to calculate an approximate 99% confidence interval for the mean weekly wage of all personnel employed in factory *A*.

The weekly wages received by a random sample of 100 personnel employed in factory *B* had a mean of £87·60 and a standard deviation of £5·20. Calculate an approximate 95% confidence interval for the difference between the mean weekly wages in the two factories. State, with your reason, whether or not your interval discredits the claim that the mean weekly wages in the two factories are equal. *(WJEC)*

22 A random sample of n observations from a population distribution had the values $x_1, x_2, \ldots, x_n$, whose mean is $\bar{x}$. Show that for any value of c,

$$\sum_{i=1}^{n} (x_i - c)^2 \equiv \sum_{i=1}^{n} (x_i - \bar{x})^2 + n(c - \bar{x})^2.$$

Hence find the value of c for which $\sum_{i=1}^{n} (x_i - c)^2$ is a minimum.

A random sample of 12 values from a normal distribution, whose mean μ and variance σ^2 are unknown, were such that

$$\sum_{i=1}^{12} x_i = 5472, \qquad \sum_{i=1}^{12} (x_i - 450)^2 = 1620.$$

   (i)   Calculate unbiased estimates of μ and σ^2.
   (ii) Determine a 95 per cent confidence interval for μ.
   (iii) Given that $(451, 463)$ was a 95 per cent confidence interval for μ based on another random sample of 12 values from the same normal distribution, deduce the corresponding unbiased estimates of μ and σ^2 from this sample. *(WJEC)*

23 a A random sample of ten observations from a normal distribution had the values $x_1, x_2, \ldots, x_{10}$ and it was found that

$$\sum_{i=1}^{10} x_i = 2.4 \qquad \text{and} \qquad \sum_{i=1}^{10} x_i^2 = 4.86.$$

Calculate 95 per cent confidence limits for the mean of the normal distribution.

   b A random sample of n observations is to be drawn from a normal distribution whose mean μ is unknown and whose standard deviation is known to be 1.5. The sample values are to be used to calculate a 95 per cent confidence interval for μ. Find the smallest n which will give a confidence interval of width less than 0.5. *(WJEC)*

24 A random sample of n observations from a population is to be used to determine confidence limits for the population mean. To justify the use of Student's t distribution for this purpose state what assumptions, if any, need to be made about
   (i) the nature of the distribution of the population,
   (ii) the value of n.

Given that $x_1, x_2, \ldots, x_{25}$ are the values of 25 random observations from a population, such that

$$\sum_{i=1}^{25} x_i = 365, \qquad \sum_{i=1}^{25} x_i^2 = 5545,$$

use t-tables to determine 95 per cent confidence limits for the population mean.

Suppose that it was known that the population standard deviation was equal to 4. Use this additional information to redetermine 95 per cent confidence limits for the population mean. *(WJEC)*

25 Explain what is meant by the sampling distribution of an estimator. Given that a random sample of n observations is drawn from a normal distribution having mean μ and variance σ^2, specify the sampling distribution of the sample mean. State what can be said about the sampling distribution of the sample mean when n is large and the parent population is not normal.

In order to estimate the mean life of an electronic component, a manufacturer carried out tests on a sample of 26 of these components. From these experiments he obtained the values $\sum x_i = 395.2$, $\sum x_i^2 = 6232$, where $x_i(i = 1 \text{ to } 26)$ is the life in hours of the ith component. Use these data to find

(i) an unbiased estimate of the mean life of the population,

(ii) an unbiased estimate of the variance of the population,

(iii) a symmetric 95% confidence interval for the mean life of the population.

If, instead, the manufacturer had tested a sample of 260 components and found $\sum x_i = 3952$ and $\sum x_i^2 = 62320$, obtain revised estimates for (ii) and (iii) above. (*JMB*)

Chapter 8

Hypothesis testing

Another important technique used in inferential statistics is called hypothesis testing. The following examples will serve as an introduction to the ideas and terms used.

8.1 Tests for a probability

Example 1
Devise a test procedure to determine whether or not a coin is unbiased.

Initially an assumption is made about the coin; for example, because the coin has a symmetrical appearance it might be assumed that the coin is unbiased. This assumption is called the *null hypothesis* and is denoted by H_0; the null hypothesis will not be rejected until there is strong evidence to the contrary. H_0 may be expressed in another way, viz. the probability p of obtaining a head on each toss is 0.5. Symbolically, this is written

$$H_0: p = 0.5.$$

A test procedure is then devised; for example, the coin may be tossed ten times and the number of heads recorded.

Next a decision rule is selected; for example, if the number of heads obtained is 0, 1, 2, 8, 9 or 10, then it may be decided that H_0 should be rejected in favour of an *alternative hypothesis*, denoted by H_1. If there is no prior reason to suppose that p is either greater than 0.5 or less than 0.5, then an appropriate alternative hypothesis is that p is not equal to 0.5. Symbolically

$$H_1: p \neq 0.5.$$

This is called a *two-sided alternative hypothesis* and the test is called a *two-tailed test*.

Finally the test is carried out and the decision reached.

There are a number of possible cases to consider.

1 If the number of heads obtained is 0, 1 or 2, H_0 would be rejected in favour of H_1 and the conclusion is that $p < 0.5$.

2 If the number of heads obtained is 3, 4, 5, 6 or 7, H_0 would be retained.

3 If the number of heads obtained is 8, 9 or 10, H_0 would be rejected in favour of H_1 and the conclusion is that $p > 0.5$.

Critical region and significance level

In the above test the number of heads X obtained in ten independent tosses of the coin is called the *test statistic*.

The set of values of the test statistic which will lead to the rejection of the null hypothesis is called the *critical region* of the test. In this case the critical region is the set $\{0, 1, 2, 8, 9, 10\}$.

It is clear that the above procedure cannot decide with certainty whether or not H_0 is true. The probability of rejecting the null hypothesis H_0 when it is true is called the *significance level* of the test and is denoted by α.

Hypothesis tests are sometimes called *significance tests* and a value of the test statistic which results in the rejection of the null hypothesis is said to be significant.

The significance level of the above test is easily calculated as follows.

If H_0 is true then $X \sim B(10, 0.5)$.

$$P(H_0 \text{ is rejected when } H_0 \text{ is true}) = P(X \leq 2) + P(X \geq 8)$$

Using **Table 1**, (page 229) with $n = 10$ and $p = 0.5$, $P(X \leq 2) = 0.055$ and, by symmetry, $P(X \geq 8) = 0.055$.

Therefore $\alpha = 0.110$, i.e. the significance level of the test is 11%.

Example 2

A psychologist has devised the following procedure as a preliminary test of the claims of persons who say they have extra-sensory perception. The psychologist deals a card at random from a standard pack and notes its colour (black or red); the subject, who cannot see the card, is then asked to write down the colour of the card; the card is replaced and the experiment repeated a further nineteen times. The number of times the subject has correctly matched the colours of the cards is noted.

a State suitable null and alternative hypotheses for the test.

b Using a significance level of approximately 5% test the claim of a subject who achieves 15 matches in the test.

a Let p be the probability of obtaining a colour match. A suitable null hypothesis is to assume that the results were obtained by guesswork.

$$H_0: p = 0.5.$$

Any low score obtained by the subject would be assigned to pure chance, since the subject would not be expected to have a value of p lower than 0.5. A suitable alternative hypothesis would be that p is greater than 0.5.

$$H_1: p > 0.5.$$

This is called a *one-sided alternative hypothesis* and the test is said to be *one-tailed*.

b Assuming that H_0 is true, X, the number of correct colour matches in 20 trials has a binomial distribution with $n = 20$ and $p = 0.5$.

Let c be the smallest value in the critical region of the test.

$$P(X \geqslant c) = 0.05$$
$$1 - P(X \leqslant c - 1) = 0.05$$
$$P(X \leqslant c - 1) = 0.95$$

From **Table 1**, (page 229) with $n = 20$ and $p = 0.5$, $P(X \leqslant 13) = 0.942$ is closest to 0.95.

$$c - 1 = 13$$
$$c = 14$$

The critical region is $\{14, 15, 16, 17, 18, 19, 20\}$ and the actual significance level is 5.8%, since $1 - 0.942 = 0.058$.

Since the value of the test statistic, 15, lies in the critical region, H_0 is rejected in favour of H_1 and the subject's claim is accepted.

An alternative method of answering part **b** does not involve finding the critical region. Instead $P(X \geqslant 15)$ is calculated and if it is less than 0.05, H_0 is rejected.

$$P(X \geqslant 15) = 1 - P(X \leqslant 14)$$

Using **Table 1**, $P(X \geqslant 15) = 1 - 0.979 = 0.021$.

Since $P(X \geqslant 15) < 0.05$, reject H_0 in favour of H_1.

Test procedure

In hypothesis testing the following procedure is carried out in the order given.

1 State the null and alternative hypotheses.
   If a definite increase (or decrease) in a population parameter is suspected use an appropriate one-sided alternative hypothesis, otherwise use a two-sided alternative hypothesis.

2 Assuming H_0 is true, determine the distribution of the test statistic.

3 Select a significance level for the test.

4 Determine the critical region.

The above stages should be completed before any sample values are considered.

5 Calculate the value of the test statistic from the sample values.

6 Determine whether or not the value of the test statistic lies in the critical region.

7 Make a conclusion concerning the population parameter.

In most examination questions all the information necessary for the solution of the question is given before the question itself is posed, thus it is quite common to give information concerning the sample values before any statement about the two hypotheses and the significance level is made. It is important, particularly in determining the appropriate alternative hypothesis, that the correct order for the test procedure should be remembered.

Exercise 8.1

1 A certain drug cures 55% of patients treated. In a clinical trial 10 patients are to be given a modified version of the drug, which is claimed to be an improvement, and the number of patients cured is to be recorded. Stating suitable null and alternative hypotheses, find the critical region of a test of the claim when the significance level is 10%. State your conclusion if 8 of the patients taking part in the trial are cured.

2 In each trial of a random experiment the probability of success is p. The null hypothesis $p = 0.4$ is to be tested against the alternative hypothesis $p \neq 0.4$ by recording the number of successes in 20 independent trials of the experiment and using a critical region of $\{0, 1, 2, 14, 15, 16, 17, 18, 19, 20\}$. Calculate the significance level of the test. State your conclusion if 16 successes are obtained in the test.

3 A certain manufacturer A of cat food claims that more than 7 cats out of 10 prefer his product to that of a rival firm B. This claim is to be tested using the following procedure: independently, each of 20 cats is to be presented with a free choice between the two brands and if at least 18 choose A's cat food then his claim will be accepted. Stating suitable null and alternative hypotheses calculate the significance level of the test.

4 A multiple-choice paper consists of 20 questions, each question having 5 alternative answers only one of which is correct. Devise a test with a significance level of approximately 5% to determine whether a particular candidate is guessing or not. Stating appropriate null and alternative hypotheses find the critical region and the actual significance level.

5 A damaged coin is to be tested to determine whether or not it is biased by tossing it 12 times and recording the number of heads obtained. Write down appropriate null and alternative hypotheses and find the critical region when the significance level is approximately 5%.

8.2 Tests for a probability using a normal approximation

Example 3

In a certain constituency 40% of the electorate voted Conservative in the last general election. An opinion poll is to be conducted in the constituency to test, using a 5% significance level, the claim that there has been a change in the level of support for the Conservative party since the last election. In the poll 263 out of 600 persons interviewed stated that they intended to vote Conservative in the next election. Write down suitable null and alternative hypotheses, carry out the test and state your conclusion.

$$H_0: p = 0.4$$

Before the opinion poll was carried out there was no indication of an increase or a decrease in the level of support for the Conservative party, so a two-sided alternative hypothesis is appropriate.

$$H_1: p \neq 0.4 \text{ (two-tailed test)}$$

Let X be the number of persons in a random sample of 600 who intend to vote Conservative in the next election. Assuming H_0 is true, $X \sim B(600, 0.4)$.

$$E[X] = np = 600 \times 0.4 = 240$$
$$V[X] = np(1 - p) = 600 \times 0.4 \times 0.6 = 144$$

Since n is large, X is approximately $N(240, 144)$.

Therefore $Z = \dfrac{X - 240}{\sqrt{144}}$ is approximately $N(0, 1)$.

$$P(X \geqslant 263) \simeq P\left(Z > \frac{262.5 - 240}{12}\right) \qquad \text{(using a continuity correction)}$$
$$= P(Z > 1.875)$$
$$= 1 - P(Z < 1.875)$$
$$= 1 - 0.9696 \qquad \text{(using \textbf{Table 3}, page 231)}$$
$$= 0.0304$$

Since $P(X \geqslant 263) > 0.025$ (two-tailed 5% test), do not reject H_0 and conclude that there has been no change in the level of support.

A variation of the above method does not require $P(X \geqslant 263)$ to be evaluated. It is only necessary to compare the value of the test statistic z with the critical values of Z for the test.

As before $Z = \dfrac{X - 240}{\sqrt{144}}$ is approximately $N(0, 1)$.

The value of the test statistic $z = \dfrac{262.5 - 240}{12} = 1.875$.

From **Table 3b**, (page 231) $P(Z > 1.96) = 0.025$ and by symmetry $P(Z < -1.96) = 0.025$. Therefore the critical values for a two-tailed 5% significance test using a standard normal statistic are ± 1.96.

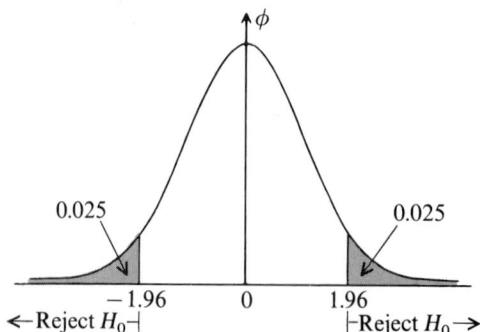

Since $z = 1.875$ is not in the critical region H_0 is not rejected.

Some critical regions for tests for a probability using a standard normal statistic

Two-tailed test with 5% significance level

$H_0: p = p_0$

$H_1: p \neq p_0$

Critical region: $(-\infty, -1.96) \cup (1.96, \infty)$

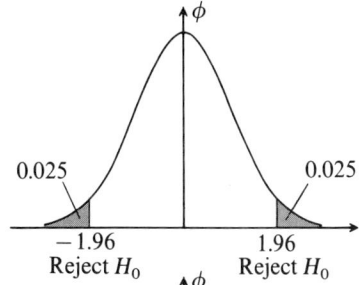

One-tailed test with 5% significance level

$H_0: p = p_0$

$H_1: p > p_0$

Critical region: $(1.64, \infty)$

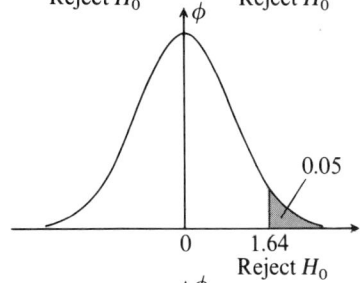

One-tailed test with 5% significance level

$H_0: p = p_0$

$H_1: p < p_0$

Critical region: $(-\infty, -1.64)$

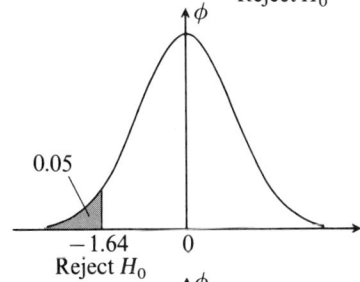

Two-tailed test with 10% significance level

$H_0: p = p_0$

$H_1: p \neq p_0$

Critical region: $(-\infty, -1.64) \cup (1.64, \infty)$

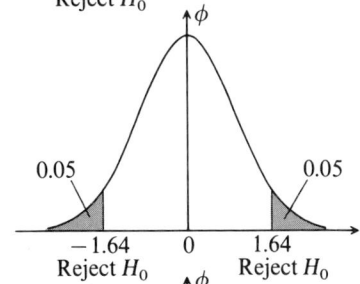

One-tailed test with 10% significance level

$H_0: p = p_0$

$H_1: p > p_0$

Critical region: $(1.28, \infty)$

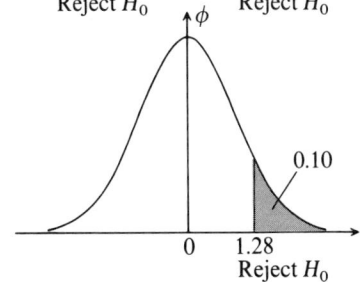

Two-tailed test with 1% significance level

$H_0: p = p_0$

$H_1: p \neq p_0$

Critical region: $(-\infty, -2.58) \cup (2.58, \infty)$

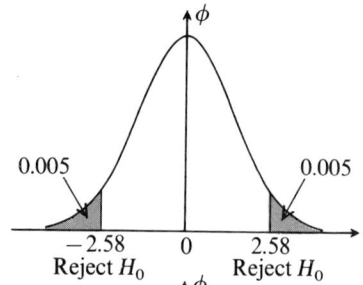

0.005   0.005

−2.58   0   2.58
Reject H_0 Reject H_0

One-tailed test with 1% significance level

$H_0: p = p_0$

$H_1: p < p_0$

Critical region: $(-\infty, -2.33)$

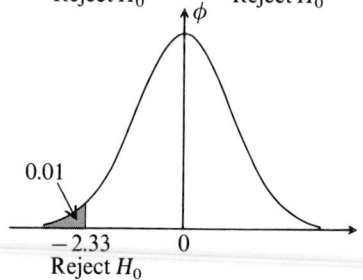

0.01

−2.33   0
Reject H_0

Exercise 8.2

1 In a certain constituency 36% of the electorate voted Labour in the last general election. After a campaign conducted by the Labour Party, an opinion poll was commissioned to test, using a 5% significance level, the claim that the level of support for the Labour Party had increased since the last election. In the poll 162 out of 400 stated that they intended to vote Labour in the next election. Write down suitable null and alternative hypotheses, carry out the test and state your conclusion.

2 In order to test whether or not a coin is unbiased, it is to be tossed 200 times and the number of heads recorded. State suitable null and alternative hypotheses. Using a 5% significance level, carry out the test and state your conclusion when the number of heads obtained is

   a   116   b   110   c   85.

3 A drug company claims that at least 75% of persons suffering from a headache gain instant relief by using its patent headache remedy. A sceptical committee decide to test this claim, using a 10% significance level. In each of the following, state the conclusion that should be drawn

   a   when 70 patients out of a random sample of 100 gained instant relief,

   b   when 140 patients out of a random sample of 200 gained instant relief.

4 The sex ratio of rabbits bred in captivity is 11 males to 9 females; to investigate whether or not the sex ratio is the same for wild rabbits, a random sample of 150 wild rabbits was obtained and it was found that this contained 66 males. Carry out a 1% significance test and state your conclusion.

5 A particular type of seed has an 80% germination rate. A seed merchant claims that after treatment the germination rate is increased. In a random sample of 500 treated seeds 422 germinated. Using a 1% significance level, test this claim.

6 A casino is accused of using an unfair roulette wheel. On each spin of the wheel there are 37 possible scores, labelled 0 to 36, and all should have equal probability. An inspector spins the wheel 740 times and the number 0 (on which the casino wins) turned up 35 times. Is this evidence sufficient to conclude that the wheel is unfair?

8.3 Tests for the mean of a normal distribution with known variance

Suppose that $X \sim N(\mu, \sigma^2)$, where σ^2 is known, and that it is required to test the null hypothesis $H_0: \mu = \mu_0$. Tests of H_0 will be based upon the observed value of $\bar{X}$, the mean of a random sample of n observations of X.

Assuming H_0 is true the distribution of $\bar{X}$ is $N\left(\mu_0, \dfrac{\sigma^2}{n}\right)$ and the distribution of

the test statistic $Z = \dfrac{\bar{X} - \mu_0}{\dfrac{\sigma}{\sqrt{n}}}$ is $N(0, 1)$. Tests having the critical regions and the

significance levels listed in the previous section may then be carried out.

Example 4

When properly adjusted, a machine filling bottles of lemonade delivers amounts which are normally distributed with mean 0.750 l and standard deviation 0.008 l. As part of a quality control process, a sample of 20 bottles was checked and found to contain a mean of 0.746 l of lemonade. Is there sufficient evidence, at the 5% significance level, to state that the machine needs adjusting?

A suitable null hypothesis is that the machine does not need adjusting.

$$H_0: \mu = 0.750.$$

Since, before the sample values are known, there is no indication or suspicion that the machine is delivering too little or too much lemonade, the alternative hypothesis should be two-sided.

$$H_1: \mu \neq 0.750.$$

The critical region of a two-tailed 5% test using a standard normal statistic is $(-\infty, -1.96) \cup (1.96, \infty)$.

Let $\bar{X}$ be the mean of a random sample of 20 bottles of lemonade.

Assuming H_0 is true, $\bar{X} \sim N\left(0.750, \dfrac{0.008^2}{20}\right)$

and $\qquad Z = \dfrac{\bar{X} - 0.750}{\dfrac{0.008}{\sqrt{20}}} \sim N(0, 1).$

For the sample obtained $\bar{x} = 0.746$.

Therefore the value of the test statistic is given by

$$z = \frac{0.746 - 0.750}{\dfrac{0.008}{\sqrt{20}}} = -2.24.$$

Since $z = -2.24$ is in the critical region, the result is significant at the 5% level and H_0 is rejected. There is sufficient evidence to suggest that the machine is delivering less than 0.750 l, on average, and therefore it needs adjusting.

Exercise 8.3

1   The times of a 400 m runner for last season were normally distributed with mean 51.62 s and standard deviation 0.75 s. In an attempt to improve his times the runner trained at altitude during the winter. In the first ten races of this season he averaged 51.18 s. Test, using a 5% significance level, whether the altitude training has resulted in improved times, on average.

2   When properly adjusted, a weighing machine produces readings which are normally distributed with mean equal to the true weight of the object being weighed and with standard deviation 3.2 mg. For calibration purposes, an object whose weight is known to be 500 mg, was weighed independently 16 times and a mean reading of 501.4 mg was calculated. Is there significant evidence, at the 5% level, to support the claim that the machine needs adjustment.

3   The fuel consumption of cars of a certain model, as measured on a standard test, is normally distributed with mean 41.2 mpg and standard deviation 1.5 mpg. A minor modification is claimed to improve the fuel consumption of the cars. Test, using a 10% significance level, this claim when a random sample of 24 of the modified cars achieved a mean fuel consumption of 41.8 mpg on the standard test.

4   The number of letters per word in poems known to have been written by a particular author A is normally distributed with mean 5.4 and standard deviation 1.6. A fragment of a poem, consisting of 51 words, has a mean word length of 5.8 letters. Is there sufficient evidence based on word length to reject the hypothesis that A is the author of the fragment? Use a 10% significance level.

5 A production line is set to produce articles of mass 454 g. The mean mass of a random sample of 12 articles from the line is found to be 452.3 g. Assuming that the masses of the articles produced are normally distributed with variance $16 \, g^2$, test, using a 1% significance level, whether the line is producing articles having a mean mass below the nominal mass.

8.4 Tests for the mean of a normal distribution with unknown variance

Suppose that $X \sim N(\mu, \sigma^2)$, where σ^2 is unknown and it is required to test the null hypothesis $H_0: \mu = \mu_0$. As in the previous section, tests of H_0 will be based on the observed value of the mean $\bar{X}$ of a random sample of n observations of X. In this case, since σ^2 is unknown, the test is based on the statistic T,

where $T = \dfrac{\bar{X} - \mu_0}{\dfrac{S}{\sqrt{n}}}$ has a t-distribution with $n - 1$ degrees of freedom.

The tests are carried out in a manner similar to that used in the previous section except that the critical values, which depend upon the sample size as well as the significance level and the alternative hypothesis, are given in **Table 4**, (page 232).

Example 5

The weights of eggs laid by a hen when fed on ordinary corn are known to be normally distributed with mean 32 g. When the hen was fed on a diet of vitamin-enriched corn a random sample of 10 eggs was weighed and the following results (in g) were recorded:

31,	33,	34,	35,	35,	36,	32,	31,	36,	37.

Test, using a 5% significance level, the claim that the new diet has increased the mean weight of eggs laid by the hen by more than 1 g.

The null hypothesis is that the new diet has increased the mean weight by 1 g.

$$H_0: \mu = 33.$$

Since the new diet is claimed to increase the mean weight of the eggs laid by more than 1 g, the alternative hypothesis is one-sided.

$$H_1: \mu > 33.$$

Using **Table 4** with $(n - 1)$ degrees of freedom, i.e. $v = 9$, $P(T > 1.83) = 0.05$.

Therefore the critical region of the one-tailed 5% test is $(1.83, \infty)$.

For the sample values $\sum x = 340$ and $\sum x^2 = 11602$.

$$\bar{x} = \frac{\sum x}{n}$$

$$\bar{x} = \frac{340}{10}$$

$$\bar{x} = 34$$

$$s^2 = \frac{\sum x^2 - n\bar{x}^2}{n-1}$$

$$s^2 = \frac{11602 - 10 \times 34^2}{9} = 4.666667$$

$$s = 2.160$$

The value of the test statistic

$$t = \frac{\bar{x} - \mu}{\dfrac{s}{\sqrt{n}}}$$

$$t = \frac{34 - 33}{\dfrac{2.160}{\sqrt{10}}} = 1.46$$

Since the value of the test statistic $t = 1.46$ does not lie in the critical region, do not reject H_0. The claim that the new diet increases the mean weight by more than 1 g is not accepted at the 5% significance level.

Exercise 8.4

1 A production line is designed to produce 120 finished articles per hour. The line foreman has reason to suspect that the rate of production has decreased. Observations over 8 separate hourly periods gave the following numbers of finished articles:

116, 122, 104, 98, 125, 114, 112, 110.

Carry out a 5% significance test of the foreman's suspicion. State any assumption that you need to make in order to carry out the test.

2 A certain drug is to be tested for its effect on diastolic blood pressure. A random sample of 11 patients had their blood pressures measured before and after the drug was administered and the results are recorded below:

Patient	A	B	C	D	E	F	G	H	I	J	K
Before	88	101	94	92	87	98	95	102	85	93	98
After	85	103	98	92	90	105	99	100	88	96	99

Test, using a 5% significance level, whether the drug has an effect on blood pressure. State any assumption that you need to make in order to carry out the test.

3 A battery manufacturer claims that, when used in a particular type of calculator, his batteries have lifetimes which are normally distributed with mean 24 hours. The lifetime x hours of each battery in a random sample of 15 was measured and the following results calculated:

$$\sum x = 351.0, \qquad \sum x^2 = 8233.56.$$

Test, using a 10% significance level, the hypothesis that the mean lifetime of the batteries differs from the manufacturer's claim.

4 Flour is sold in 1.5 kg packs. A sample of 21 packs is selected at random and the mass x kg of each recorded. The following results were calculated:

$$\sum x = 31.08, \qquad \sum x^2 = 46.0704.$$

Assuming that the masses of the packs are normally distributed, test at the 5% level whether the packs are significantly underweight.

5 The manufacturer of steel rods claims that the mean breaking strength is 1000 N. Five rods are tested to destruction and the breaking strengths, in N, were

912,   945,   802,   957,   984.

Assuming that the breaking strengths are normally distributed and using a 1% significance level, test the manufacturer's claim against the alternative hypothesis that the mean breaking strength is less than 1000 N.

8.5 Tests for the difference between the means of two distributions

Suppose that X_1 and X_2 are two independent random variables such that $X_1 \sim N(\mu_1, \sigma_1{}^2)$, $X_2 \sim N(\mu_2, \sigma_2{}^2)$ and $\sigma_1{}^2$, $\sigma_2{}^2$ are both known. Tests of the null hypothesis $H_0: \mu_1 - \mu_2 = 0$ (or any other specified value) will be based on the observed value of $\bar{X}_1 - \bar{X}_2$, where $\bar{X}_1$ denotes the mean of a random sample of n_1 observations of X_1, and $\bar{X}_2$ denotes the mean of an independent random sample of n_2 observations of X_2. As shown on page 91, the sampling distribution of $\bar{X}_1 - \bar{X}_2$ is normal with mean $\mu_1 - \mu_2$ and variance $\dfrac{\sigma_1{}^2}{n_1} + \dfrac{\sigma_2{}^2}{n_2}$.

The test statistic is

$$Z = \frac{(\bar{X}_1 - \bar{X}_2) - (\mu_1 - \mu_2)}{\text{SE}[\bar{X}_1 - \bar{X}_2]}, \text{where SE}[\bar{X}_1 - \bar{X}_2] = \sqrt{\frac{\sigma_1{}^2}{n_1} + \frac{\sigma_2{}^2}{n_2}}.$$

Z has a $N(0, 1)$ distribution and tests of H_0, which use the observed value of Z together with the critical regions and the significance levels given in **8.2**, may be used.

STATISTICS 2

Even when·the distributions of X_1 and X_2 are not normal the tests may still be used, provided n_1 and n_2 are both large, as, by the Central Limit Theorem, the distributions of $\bar{X}_1$ and $\bar{X}_2$ are approximately normal. Furthermore, when σ_1^2 and σ_2^2 are unknown, they may be replaced by s_1^2 and s_2^2, their sample unbiased estimates, without unduly affecting the significance level of the tests.

When n_1 and n_2 are both large the test statistic is

$$Z = \frac{(\bar{X}_1 - \bar{X}_2) - (\mu_1 - \mu_2)}{\text{ESE}[\bar{X}_1 - \bar{X}_2]}, \text{ where } \text{ESE}[\bar{X}_1 - \bar{X}_2] = \sqrt{\frac{s_1^2}{n_1} + \frac{s_2^2}{n_2}}.$$

Z has an approximate $N(0, 1)$ distribution and tests of H_0 may be based on its observed value.

Example 6

A researcher wishes to test, using a 5% significance level, the null hypothesis that the mean height of men is 15 cm more than the mean height of women against the alternative hypothesis that the mean height of men is less than 15 cm more than the mean height of women. From a random sample of 100 men unbiased estimates of 178.1 cm and 25.12 cm² for the mean and the variance, respectively, of the height of men were calculated. An independent random sample of 100 women gave unbiased estimates of 164.2 cm and 15.91 cm² for the mean and the variance, respectively, of the height of women. Carry out the test and state your conclusion.

$H_0: \mu_M - \mu_W = 15$

$H_1: \mu_M - \mu_W < 15$ (one-tailed)

The critical region of this one-tailed 5% test using a standard normal statistic is $(-\infty, -1.64)$

Assuming H_0 is true, and since n_M and n_W are both large, the distribution of $(\bar{X}_M - \bar{X}_W)$ is approximately normal with mean 15.

$$\text{ESE}[\bar{X}_M - \bar{X}_W] = \sqrt{\frac{s_M^2}{n_M} + \frac{s_W^2}{n_W}}$$

$$= \sqrt{\frac{25.12}{100} + \frac{15.91}{100}}$$

$$= 0.6405466$$

$$Z = \frac{(\bar{X}_M - \bar{X}_W) - (\mu_M - \mu_W)}{\text{ESE}[\bar{X}_M - \bar{X}_W]} \sim N(0, 1)$$

$$z = \frac{(178.1 - 164.2) - 15}{0.6405466}$$

$$z = -1.72$$

Since $z = -1.72$ is in the critical region $(-\infty, -1.64)$ the result is significant at the 5% level. Reject H_0 in favour of H_1 and conclude that the mean height of men is less than 15 cm more than the mean height of women.

Exercise 8.5

1   In a butter-packing factory the quantity of butter packed in a day using a certain type of machine is normally distributed with standard deviation 43 kg. Two packers average 2186 kg and 2162 kg per day over a 20-day month. Are the performances of the two packers significantly different at the 5% level?

2   An oil company has produced an additive for petrol which, it claims, will increase the number of miles per gallon of cars. To test this claim a consumer organisation tests 100 cars with the additive and 120 cars without the additive. Letting x_i ($i = 1$ to 100) be the numbers of miles per gallon obtained by the first 100 cars and y_j ($j = 1$ to 120) the numbers of miles per gallon of the other 120 cars, the following results were obtained:

$$\sum x_i = 4230 \quad \sum x_i^2 = 179820 \quad \sum y_j = 5016 \quad \sum y_j^2 = 212048$$

Using a 5% significance level, test the oil company's claim and state your conclusion.

3   In one Telecom area it was found over a certain period that 400 customers, randomly selected, made a total of 43261 calls. During the same period, a random sample of 500 customers in another area made a total of 56076 calls. Assuming that the standard deviation of the number of calls made by customers during the period was 12 in both areas, test, using a 1% significance level, the claim that the mean number of calls per customer in the second area exceeds that in the first area by more than 2.

4   In an investigation into the effectiveness of a particular course in speed reading, a group of 500 students was split at random into two equal groups, A and B. The students in group A were given the course in speed reading, while the students in group B were given no special instruction. At the end of the course each student was asked to read the same passage and the time taken was measured. The mean and the variance of the times taken for each group were calculated.

Group A: mean 139.2 s   variance $10\,\text{s}^2$
Group B: mean 144.7 s   variance $20\,\text{s}^2$

Stating appropriate null and alternative hypotheses, test, using a 10% significance level, whether the course has been effective in reducing the mean reading time for the passage by more than 5 s.

5 In order to test which of two methods, A and B, is the quicker in finishing a particular task, two groups of workers were chosen at random; the first group consisting of 78 workers used method A and the second group of 92 workers used method B. The sum of the times taken and the sum of the squares of the times taken were calculated for each group.

First group: 998.4 min   12890.40 min^2
Second group: 1122.4 min   14308.44 min^2

Stating suitable null and alternative hypotheses carry out the test using a 10% significance level and state your conclusion.

8.6 Tests for the mean of a Poisson distribution

Suppose X has a Poisson distribution with mean α and that it is required to test the null hypothesis H_0; $\alpha = \alpha_0$. Tests of H_0 will be based on Y, the sum of n independent observations of X: as shown in **3.3** on page 44, Y has a Poisson distribution with mean $n\alpha$. There are two cases to consider:

1 When $n\alpha$ is small, in which case **Table 2**, (page 230) may be useful.
2 When $n\alpha$ is large and a normal approximation to the distribution of Y will be appropriate.

Example 7

The number of new cars sold per week in a small garage has a Poisson distribution with mean 1.5. The owners of the garage, wishing to investigate the effect of the appointment of a new salesman, decide to use the number of sales of new cars in a 4-week period as a test statistic. Carry out the test, using a 5% significance level, when the number of sales recorded in the 4-week period after the appointment of the new salesman was 11.

Let X_i be the number of cars sold in the ith week after the appointment of the salesman. Assume that the X_i are independent and that each has a Poisson distribution with mean α.

If $Y = X_1 + X_2 + X_3 + X_4$, then $Y \sim Po(4\alpha)$.

A suitable null hypothesis H_0 is that the new salesman has had no effect on the mean number of cars sold in a week.

$$H_0: \alpha = 1.5.$$

Since the appointment of the salesman may cause an increase or a decrease in the mean number of cars sold in a week, the alternative hypothesis is two-sided.

$$H_1: \alpha \neq 1.5.$$

Assuming that H_0 is true, $Y \sim Po(6)$ and the probability that Y is greater than or equal to the observed value 11 may be calculated.

$$P(Y \geq 11) = 1 - P(Y \leq 10)$$
$$= 1 - 0.957 \qquad \textbf{(Table 2)}$$
$$= 0.043$$

Since $P(Y \geq 11) = 0.043 > 0.025$ (two-tailed 5% test), the result is not significant at the 5% significance level and H_0 is not rejected.

It may be noted that, since Y is a discrete random variable, it is not possible to carry out the test with a significance level of exactly 5%. Referring to **Table 2**, $P(Y \leq 1) = 0.017$ and $P(Y \geq 12) = 0.020$, so a test with critical region $\{0, 1, 12, 13, 14, \ldots\}$ would have a significance level of 0.037 (or 3.7%), which is the nearest to 5% for a two-tailed test without exceeding 5%.

Example 8

As a result of the test carried out in **Example 7** the owners of the garage suspect that the salesman may have improved the level of sales, so they decide to apply another test based on the number of new cars sold in the next 20-week period. Carry out the test, using a 5% significance level, when the number of new cars sold in the 20-week period was 41.

If $Y = X_1 + X_2 + \cdots + X_{20}$ then $Y \sim Po(20\alpha)$.

As before $H_0: \alpha = 1.5$.

Since it is suspected that the salesman has improved the level of sales the alternative hypothesis is one-sided.

$$H_1: \alpha > 1.5 \text{ (one-tailed)}$$

Assuming that H_0 is true, $Y \sim Po(30)$ and $Z = \dfrac{Y - 30}{\sqrt{30}}$ is approximately $N(0, 1)$.

For the observed value of $y = 41$ the value of the test statistic z is given by

$$z = \frac{40.5 - 30}{\sqrt{30}} = 1.92 \, (2 \, \text{d.p.}).$$

Since $z = 1.92 > 1.64$ the result is significant at the 5% level. H_0 is rejected in favour of H_1 and the conclusion is that the salesman has increased the level of sales, on average.

Exercise 8.6

1   The number of claims per week made to the office of a life insurance company has a Poisson distribution with mean 1.25. In a 4-week period 10 claims were made. Is there enough evidence at the 5% significance level to suggest that the mean number of claims per week has increased?

2   The number of errors per page made by a typist has a Poisson distribution with mean 1.4. After some training the typist was given 5 pages to type and made 3 errors. Test, using a 10% significance level, the claim that the training has reduced the mean number of errors per page made by the typist.

3 The number of copies of a certain magazine sold per week in a particular shop has a Poisson distribution with mean 5. The magazine was advertised on television during an 8-week period and the number of copies sold in the shop was 48. Test, using a 5% significance level, the claim that the television-advertising campaign has increased the mean number of copies sold per week in the shop.

4 A 500 m roll of cloth is considered to be acceptable provided that the mean number of flaws per metre length is not more than 0.2. An inspection of a 10 m length of the roll revealed 6 flaws. Assuming that the number of flaws per metre length has a Poisson distribution, use a 1% significance level to test whether the roll is acceptable.

5 The number of accidents occurring per week along a certain stretch of road has a Poisson distribution with mean 4. In a 25-week period after some alterations a total of 82 accidents were recorded. Is there enough evidence at the 5% significance level to support the claim that the mean number of accidents per week along the road has been altered?

8.7 Type-1 and type-2 errors

When any hypothesis test is applied there are two possible types of error which may arise:

1 a *type-1 error* occurs if H_0 is rejected when, in fact, it is true,

2 a *type-2 error* occurs if H_0 is not rejected when, in fact, it is false.

The probability of a type-1 error occurring is denoted by α. The probability of a type-2 error occurring is denoted by β; it can only be calculated when a particular value of the parameter out of those specified by H_1 is stipulated.

Example 9

X is a normally distributed random variable having mean μ and standard deviation 20. The null hypothesis $H_0: \mu = 100$ is to be tested against the alternative hypothesis $H_1: \mu > 100$, using a 2.5% significance level. The mean $\bar{X}$ of a random sample of 16 observations of X is to be used as the test statistic. Find the critical region of the test and deduce the probability of making a type-2 error when, in fact, $\mu = 115$.

Assuming H_0 is true, $\bar{X} \sim N\left(100, \dfrac{20^2}{16}\right)$ i.e. $\bar{X} \sim N(100, 25)$.

Therefore $Z = \dfrac{\bar{X} - 100}{5} \sim N(0, 1)$.

The critical value of Z for a one-tailed 2.5% significance test is 1.96, so the critical value, c, of $\bar{X}$ for the test is given by

$$\frac{c - 100}{5} = 1.96$$

$$c = 109.8.$$

The critical region is $(109.8, \infty)$

P(type-2 error) = P(H_0 is not rejected when H_1 is true).

Assuming H_1 is true, $\bar{X} \sim N(115, 25)$.

Therefore $Z = \dfrac{\bar{X} - 115}{5} \sim N(0, 1)$

$$\beta = P(\bar{X} < 109.8 | H_1 \text{ is true})$$

$$= P\left(Z < \frac{109.8 - 115}{5}\right)$$

$$= P(Z < -1.04)$$

$$= 1 - 0.8508 \qquad (\textbf{Table 3}, \text{ page } 231)$$

$$= 0.1492$$

Therefore, when $\mu = 115$, the probability of a type-2 error is 0.1492.

If the same question is repeated using 10%, 5%, 1% and 0.5% significance levels the resulting values for β are given in the following table.

α	0.10	0.05	0.025	0.01	0.005
β	0.043	0.088	0.149	0.250	0.336

It should be noted that as α decreases β increases and vice versa. It is desirable that both errors should be as small as possible but it is impossible (for a fixed sample size) to reduce both simultaneously as a decrease in one results in an increase in the other.

In industrial statistics a type-1 error is often called the producer's risk because the rejection of H_0 when true means that manufactured articles meeting specifications will be rejected as sub-standard. A type-2 error is called the consumer's risk as the consumer will be sold sub-standard articles at full price.

Power of a test

The *power* of a hypothesis test is the probability of accepting H_1 when H_1 is true.

Therefore the power = $1 - \beta$ e.g. in **Example 9** the power of the test when $\mu = 115$ is $1 - 0.1492 = 0.8508$.

Clearly the power of the test depends upon the actual value of the unknown population parameter μ, but, as in **Example 9**, it may be evaluated for any arbitrarily chosen value for μ. The function, whose domain is the set of possible values for μ and whose range is the set of the corresponding powers, is called the *power function* of the test.

The powers of the test in **Example 9** have been calculated for some values of μ and the results recorded in the following table.

μ	100	105	107	110	112	115	120
Power	0.025	0.169	0.288	0.516	0.670	0.851	0.979

The graph of the power function is sketched below.

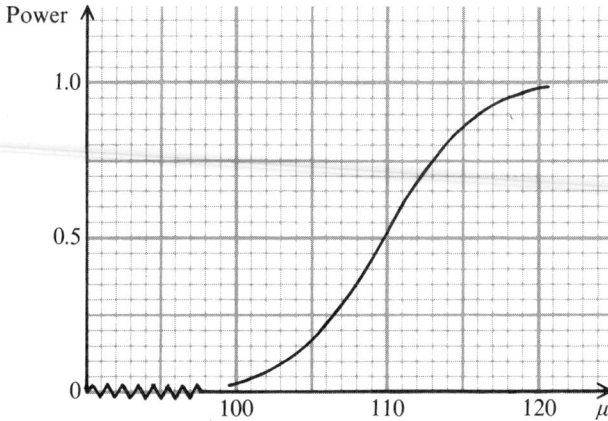

An ideal power function would take the value 0 for values of μ consistent with H_0 and the value 1 otherwise; this is not attainable in practice, but the closer the actual power function is to the ideal, the better (more powerful) the test. Increasing the size of the sample used in a test usually improves the characteristics of the power function. This is illustrated by the following diagram which compares the graphs of the power functions of the test described in **Example 9** when the sample sizes are 16, 64 and 400.

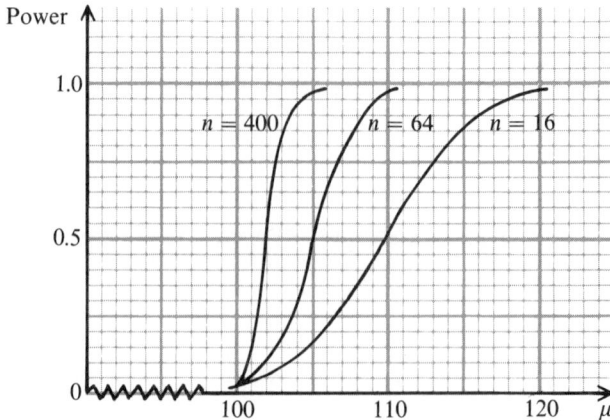

Exercise 8.7

1 X is a normally distributed random variable having mean μ and standard deviation 8. The null hypothesis H_0: $\mu = 20$ is to be tested against the alternative hypothesis H_1: $\mu > 20$ using a 5% significance level. The mean $\bar{X}$ of a random sample of 100 observations of X is to be used as the test statistic. Find the critical region of the test and calculate its power when, in fact, μ is 23.

2 $X \sim N(\mu, 6^2)$. H_0: $\mu = 44$ is to be tested against H_1: $\mu \neq 44$ using a 5% significance level. The mean $\bar{X}$ of a random sample of 25 observations of X is to be used as the test statistic. Find the critical region of the test and its power when $\mu = 43$.

3 X is a Poisson variable with mean α. H_0: $\alpha = 0.5$ is to be tested against H_1: $\alpha > 0.5$ by taking a random sample of 10 observations of X and rejecting H_0 when the sum of these observations is 9 or more. Determine the significance level of the test and the probability of making a type-2 error when $\alpha = 0.8$.

4 $X \sim Po(\alpha)$. H_0: $\alpha = 0.5$ is to be tested against H_1: $\alpha > 0.5$ by taking a random sample of 100 observations of X and rejecting H_0 when the sum of these observations is at least 60. Find the significance level of the test and the probability of making a type-2 error when $\alpha = 0.8$.

5 The probability of obtaining a head on a toss of a coin is p. H_0: $p = 0.5$ is to be tested against H_1: $p > 0.5$ by tossing the coin 20 times and rejecting H_0 when the number of heads obtained exceeds 13. Calculate the significance level of the test and its power when $p = 0.6$.

Miscellaneous Exercise 8

1 The number, X, of television sets sold in a shop per week has a Poisson distribution with mean a. Given that

$$P(X = 9) = P(X = 10), \text{ show that } a = 10.$$

In the week immediately following the appointment of a new manager the number of sets sold in the shop was 4. Stating suitable null and alternative hypotheses, test, using a 5 per cent significance level, whether the new manager has had any effect on the sales of television sets in the shop.   (*JMB*)

2 Items manufactured by process A have lengths which are normally distributed with mean μ cm and standard deviation 1 cm. A random sample of 16 items manufactured by process A had a mean length of 5.25 cm.
(i)   Calculate a 95% confidence interval for μ.
(ii) Deduce a 95% confidence interval for the proportion of items manufactured by process A that are shorter than 5 cm.
Items manufactured by process B have lengths which are normally distributed with mean λ cm and standard deviation 1 cm. A random sample of 9 items manufactured by process B had a mean length of 4.5 cm. Using a significance level of 5%, test the hypothesis that $\lambda = \mu$ against the alternative hypothesis that $\lambda \neq \mu$.   (*JMB*)

3 The tar yields in cigarettes of a certain brand are distributed normally with mean μ mg and standard deviation 0.8 mg. In a random sample of 10 cigarettes of this brand the tar yields in mg were:

17.1,	18.3,	18.9,	17.8,	16.9,
19.2,	18.2,	17.8,	18.3,	18.5.

Use a 1% significance level to test the null hypothesis that $\mu = 17.5$ against the alternative hypothesis that $\mu > 17.5$. (*JMB*)

4 For each plant in a random sample of 200 potato plants grown by method A, the weight x kg of potatoes produced was measured. The following results were calculated:

$$\sum x = 2514, \qquad \sum x^2 = 31977.$$

For each plant in a random sample of 250 potato plants grown by method B, the weight y kg of potatoes produced was measured. The following results were calculated:

$$\sum y = 3410, \qquad \sum y^2 = 47309.$$

Calculate, to two decimal places, unbiased estimates for the mean and the variance of the weights of potatoes produced per plant by each of the methods A and B.

It is claimed that the mean weight of potatoes per plant produced by method B is over 1 kg more than the mean weight of potatoes per plant produced by method A.

Stating suitable null and alternative hypotheses, test, using a 10 per cent significance level, whether the above results support this claim. (*JMB*)

5 One hundred fibres were selected at random from a very large bundle of fibres. The breaking strength x of each one was measured and the following results were calculated:

$$\sum x = 1520.0, \qquad \sum x^2 = 23112.91.$$

Find unbiased estimates for the mean μ and the variance σ^2 of the breaking strengths of fibres in the bundle. Use these results to calculate an approximate 90 per cent symmetric confidence interval for μ. Name the theorem which you have assumed.

A second random sample of one hundred fibres was taken after the fibres in the bundle had been treated with a certain chemical. The unbiased estimates of the mean μ_1 and the variance σ_1^2, calculated from this sample, were 15.28 and 0.1225, respectively. Stating suitable null and alternative hypotheses, determine whether there is sufficient evidence at the 5 per cent significance level that the treatment has increased the mean breaking strength of fibres in the bundle. (*JMB*)

6 For each packet in a random sample of 400 packets of biscuits produced by a manufacturer A the mass x grams was measured. The following results were obtained:

$$\sum x = 100\,760, \qquad \sum (x - \bar{x})^2 = 14\,364.$$

Calculate unbiased estimates of the mean and variance of the masses of all packets of biscuits produced by A. Determine an approximate 90 per cent symmetric confidence interval for the mean mass of all packets produced by A.

A random sample of 100 packets of biscuits produced by manufacturer B gave unbiased estimates of 251 and 16, respectively, for the mean and variance of the masses of packets produced by B. Determine whether there is significant evidence at the 5 per cent level to support the hypothesis that the mean mass of all packets produced by A differs from the mean mass of all packets produced by B.

Determine also whether there is significant evidence at the 5 per cent level to support the hypothesis that the mean mass of all packets produced by A exceeds the mean mass of all packets produced by B.

(JMB)

7 Independently for each month, the number of accidents per month that occur in the building industry in a certain area has a Poisson distribution with mean 1.5.
   (i) Find the probability that there will be just one accident in a month.
   (ii) Find the smallest integer m for which the probability of m or more accidents occurring in a month is less than 0.05.
   (iii) Find the largest integer n for which the probability of no accident occurring in a period of n months is at least 0.001.
   (iv) Find, in terms of e, the probability that there will be no more than one accident in a period of two months.
   (v) During a period of six months when special safety precautions had been introduced, there were exactly four accidents. Assuming that before the introduction of safety precautions the numbers of accidents during periods of six months have a Poisson distribution with mean 9.2, use a 5% significance level test to determine whether the given evidence is strong enough to conclude that the safety precautions were effective in reducing the accident rate. (JMB)

8 A random variable X has unknown mean μ and unknown variance σ^2. A random sample of 50 observations of X has sum 2625 and sum of squares 137 950. Calculate an unbiased estimate of μ and an unbiased estimate of σ^2.

Determine an approximate 95 per cent symmetrical confidence interval for μ. Test the hypothesis that $\mu = 52$ against the alternative hypothesis that $\mu \neq 52$,
   (i) at the 5 per cent level of significance,
   (ii) at the 2 per cent level of significance.

A random sample of 100 observations on another random variable Y has sum 5310 and sum of squares 285 000. Test, at the 5 per cent level of significance, whether or not the distribution means of X and Y are equal.

(JMB)

9 It is required to test the null hypothesis $H_0: \mu = 5$ against the alternative hypothesis $H_1: \mu > 5$, where μ is the mean of a normal distribution whose standard deviation is 1.6. The test procedure is to take a random sample of n observations from the normal distribution and to reject H_0 only if the sample mean exceeds c, where c is a constant.

   a In the case when $n = 16$ and $c = 5.8$,

     (i) determine the probability of a type-1 error,

     (ii) show that when $\mu = 6$, the probability of a type-2 error is approximately 0.31.

   b Suppose it is required that the probability of a type-1 error should be approximately 0.025, and that when $\mu = 6$ the probability of a type-2 error should be approximately 0.05. Show that in order to meet the first of these requirements, n and c must be chosen so that

$$c \simeq 5 + \frac{3.136}{\sqrt{n}}.$$

Find a similar relation between n and c in order to meet the second requirement. Hence determine an appropriate value of n and the corresponding value of c to two decimal places in order to meet both requirements. *(JMB)*

10 The pulse rates of men are normally distributed with mean μ and standard deviation 9. The null hypothesis $\mu = 72$ is to be tested against the alternative hypothesis $\mu \neq 72$ by finding the mean pulse rate of a random sample of 100 men. Given that the significance level of the test is to be 5 per cent, find

   (i) the critical region of the test,

   (ii) the probability of making a type-2 error when the value of μ is 69. State the power of the test in this case. *(JMB)*

11 Let p denote the probability of obtaining a head when a certain coin is tossed. To test the null hypothesis that $p = \frac{1}{2}$ against the alternative hypothesis that $p \neq \frac{1}{2}$, it is decided to toss the coin 20 times and to reject the null hypothesis if the number of heads obtained is less than 6 or more than 14. With the aid of the tables provided, find the type-1 error probability of this test procedure. Find also the type-2 error probability when $p = \frac{1}{4}$. *(JMB)*

12 Explain what is meant by a 'Type 1 error' and a 'Type 2 error' in hypothesis testing.

A manufacturer of the cat food, Mush, claims that cats prefer his product to another well known brand. This claim is to be investigated using the following procedure: each of 20 cats is to be presented independently with a free choice between the two brands and if at least 15 of the cats choose Mush then the manufacturer's claim will be accepted. Writing down appropriate null and alternative hypotheses, calculate the probability of making a Type 1 error.

If, in fact, the probability is 0.7 that a cat selected at random will choose Mush in preference to the other brand, calculate the probability that the manufacturer's claim will not be accepted. (*JMB*)

13 In each of n independent trials, the probability of success is p. Show that the probability of k successes is given by the binomial probability function

$$\binom{n}{k} p^k (1-p)^{n-k}, \qquad k = 0, 1, \ldots, n.$$

When A and B play table tennis, the probability that A wins any game is 0.6. Find, to three decimal places, the probability that A wins at least three games out of six.

C and D play ten games of which C wins eight and he then asserts that this confirms that he is the better player. With the help of the table of cumulative binomial probabilities, test, at the five per cent significance level, the null hypothesis that the probability of C winning any game is $\frac{1}{2}$, against the alternative hypothesis that it is greater than $\frac{1}{2}$.

E and F play 25 games, of which E wins 19. Use the normal approximation to the binomial distribution to test, at the five per cent level, whether this provides significant evidence of E's superiority over F at table tennis. (*JMB*)

14 It is suspected that a particular coin is biased. In each of the following cases use a 10% significance level to test the hypothesis that the coin is unbiased against the alternative hypothesis that it is biased.

(i) In 10 tosses of the coin a head is obtained twice.

(ii) In 100 tosses of the coin a head is obtained 59 times. (*JMB*)

15 Experiments were conducted to determine whether an antibiotic spray is effective in reducing the number of bacterial colonies that will develop in dishes of nutrient exposed to an infected environment. It is known that the mean number of colonies that will develop per unsprayed dish is 7.5 and that, independently for each sprayed dish, the number of colonies that will develop has a Poisson distribution.

(i) In an experiment in which one dish of the nutrient was sprayed the number of colonies that developed was 3. Determine whether this result provides significant evidence at the 5% level that the spray is effective in reducing the mean number of bacterial colonies that develop.

(ii) In another experiment, four dishes were sprayed and the numbers of colonies that developed were 3, 5, 6 and 4, respectively. Given that the sum of independent Poisson random variables is also a Poisson random variable, determine whether these results provide significant evidence at the 5% level that the spray is effective. (*JMB*)

16 Explain what is meant by the sampling distribution of a sample mean.

Let $\bar{X}$ denote the mean of a random sample of m observations from a distribution having mean μ_1 and variance $\sigma_1{}^2$, and let $\bar{Y}$ denote the mean of an independent random sample of n observations from a distribution having mean μ_2 and variance $\sigma_2{}^2$. Given that both m and n are large, state the approximate probability distribution of $\bar{X} - \bar{Y}$. Explain how $\sigma_1{}^2$ and $\sigma_2{}^2$ may be estimated from the samples when their actual values are not known.

The sum of the lengths of 100 articles manufactured in one section of a factory is 305 cm and the sum of their squares is 1225 cm². The sum of the lengths and sum of squares of the lengths of 50 such articles manufactured in a different section are 180 cm and 746 cm². Derive an approximate 95 per cent symmetrical confidence interval for the difference between the two distribution means.

Test, at the 5 per cent level of significance, the hypotheses that

(i) the mean length of articles manufactured in the first section of the factory is 3 cm;

(ii) the difference between the means of the lengths of articles manufactured in the two sections of the factory is zero. (*JMB*)

17 A continuous random variable X has the rectangular distribution over the interval $(-\theta, 2)$, where θ is an unknown positive constant. Find, in terms of θ, the probability that a randomly observed value of X will be negative.

In a random sample of 20 observed values of X it was found that 17 were negative and 3 were positive. Use tables to test, at the 2% significance level, the hypothesis that $\theta = 3$ against the alternative hypothesis that $\theta > 3$. (*JMB*)

18 Two normal distributions have means μ and λ, and have common standard deviation 5. The null hypothesis

$$H_0: \mu - \lambda = -2$$

is to be tested against the alternative hypothesis

$$H_1: \mu - \lambda = 2.$$

For this purpose it is decided to take independent random samples of size n from each of the two distributions and to reject H_0 only if $\bar{x} > \bar{y}$, where $\bar{x}$ and $\bar{y}$ are the means of the samples from the distributions having means μ and λ, respectively. Show that for this decision rule, the probabilities of a type-1 error and a type-2 error are equal.

Find the smallest value of n for each of these error probabilities to be less than 0.05. (*JMB*)

140

19 The probability density function f of the continuous random variable X is such that

$$f(x) = \alpha x^{-(\alpha+1)}, x \geq 1,$$

$$f(x) = 0, \text{ otherwise,}$$

where α is an unknown positive constant.

  (i) Determine the distribution function of X. For the case when $\alpha = 2$, find the median value of X.

 (ii) To test the null hypothesis that $\alpha = 2$ against the alternative hypothesis that $\alpha \neq 2$, it is decided to take a random sample of 100 observations of X and to count the number, R, that have values greater than 5. Given that the chosen significance level is 5% and that the observed value of R is 9, use an appropriate approximation to the sampling distribution of R when $\alpha = 2$ to carry out the test. State whether your conclusion is that $\alpha = 2$, $\alpha < 2$ or $\alpha > 2$.   (*JMB*)

20 The random variables X and Y are known to be normally distributed with means μ and λ, and variances 4 and 16, respectively. A test is required of the null hypothesis that $\mu = \lambda$ against the alternative hypothesis that $\mu \neq \lambda$. The test is to be based on the means of independent random samples of 16 observations of X and 25 observations of Y.

  (i) One possible test is to reject the null hypothesis only if the *numerical* difference between the two sample means exceeds a certain value c. Obtain the value of c for this test to have a significance level of 0.05.

 (ii) Another possible test is to reject the null hypothesis only if the 95% symmetrical confidence intervals for μ and λ do not overlap. Show that the significance level of this test is approximately 0.007.   (*JMB*)

Chapter 9

Linear relationships

In physics, chemistry and other branches of science there are many instances
of two quantities which are linearly related. In order to estimate the precise
form of the relationship, experiments are performed and data, in the form of
pairs of observations (x_1, y_1), (x_2, y_2), ..., (x_n, y_n), are collected. These data then
form the basis from which the estimate is derived. One easy method of
achieving this involves plotting the pairs of values (x_i, y_i) on a scatter diagram
and then a line of 'best' fit is drawn by eye.

Scatter diagram

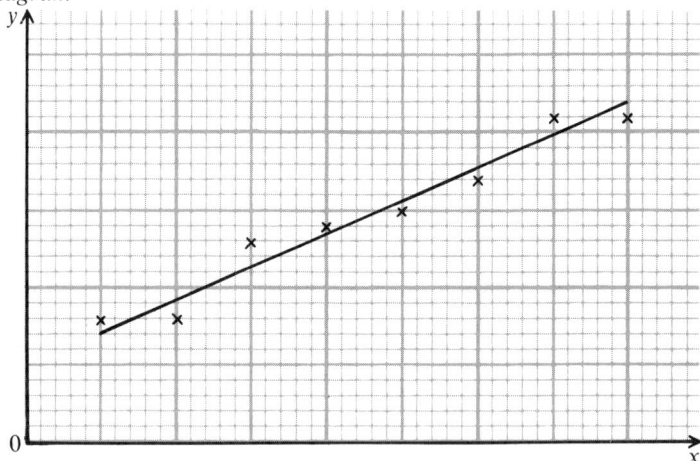

Clearly this method has many disadvantages. Since the method is subjective,
no two persons will agree precisely on a line of 'best' fit and it will not be
possible to make any probabilistic statements about the closeness of the fitted
line to the true line. A more formal method, having neither of the above
disadvantages, is called *the method of least squares*. Before introducing this
method, the reader is advised to recall the following results:

1 $\displaystyle\sum_{i=1}^{n} k = nk,$

2 $\displaystyle\sum_{i=1}^{n} kx_i = k \sum_{i=1}^{n} x_i,$

3 $\displaystyle\sum_{i=1}^{n} (x_i + y_i) = \sum_{i=1}^{n} x_i + \sum_{i=1}^{n} y_i,$

4 $\displaystyle \bar{x} = \frac{\sum_{i=1}^{n} x_i}{n}$ or $\displaystyle\sum_{i=1}^{n} x_i = n\bar{x}.$

These results were established on pages 20 and 21 of Book 1.

9.1 Method of least squares

Suppose that (x_1, y_1), (x_2, y_2), ..., (x_n, y_n) are n experimentally obtained pairs of values of x and y, where it is known that x and y are linearly related. It will be assumed that the values $x_1, x_2, ..., x_n$ are accurate and that the observed values $y_1, y_2, ..., y_n$ are subject to measurement error.

Least squares estimate of a linear relationship of the form y = mx

Initially, in order to simplify the working, the linear relationship between x and y will be assumed to take the form $y = mx$, where m is an unknown constant.

For each pair of observations (x_i, y_i)

$$y_i = mx_i + e_i, \qquad i = 1, 2, ..., n,$$

where $\quad e_i = y_i - mx_i$ is the error in the observed value of y when $x = x_i$.

Suppose that $y = \hat{m}x$ is an estimate of the line $y = mx$ and that the discrepancy between the observed value y_i and $\hat{m}x_i$ is d_i. Then

$$d_i = y_i - \hat{m}x_i, \qquad i = 1, 2, ..., n.$$

The method of least squares searches for the value of $\hat{m}$ which will minimise S, the sum of the squares of these discrepancies.

Scatter diagram

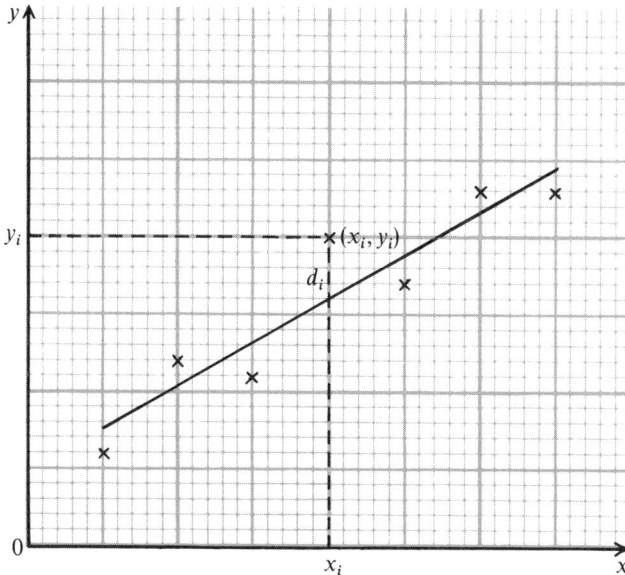

STATISTICS 2

In what follows it is important to remember that S and $\hat{m}$ are the variables and that the values x_i and y_i ($i = 1, 2, \ldots, n$) are known.

$$S = \sum (y_i - \hat{m}x_i)^2$$

Differentiate w.r.t. $\hat{m}$:

$$\frac{dS}{d\hat{m}} = \sum 2(y_i - \hat{m}x_i)(-x_i)$$

$$\frac{dS}{d\hat{m}} = 2\sum (-x_iy_i + \hat{m}x_i^2)$$

$$\frac{dS}{d\hat{m}} = 2\left[-\sum x_iy_i + \hat{m}\sum x_i^2 \right] \qquad (1)$$

For stationary values of S, $\dfrac{dS}{d\hat{m}} = 0$, which gives

$$\hat{m} = \frac{\sum x_iy_i}{\sum x_i^2}.$$

Differentiating (1) w.r.t. $\hat{m}$ gives

$$\frac{d^2S}{d\hat{m}^2} = 2\sum x_i^2.$$

Since $2\sum x_i^2 > 0$, it follows that the stationary value of S is a minimum.

The line $y = \hat{m}x$, where

$$\hat{m} = \frac{\sum x_iy_i}{\sum x_i^2},$$

is called *the least squares estimate of the line* $y = mx$.

Example 1

In Physics, Hooke's Law states that the extension of an elastic string is directly proportional to the tension in the string. In an experiment to estimate the precise relationship for a particular string, the extension y cm was measured for each of ten controlled values of the tension x N. The results obtained are recorded below.

x	1.0	1.5	2.0	2.5	3.0	3.5	4.0	4.5	5.0	5.5
y	4.1	6.2	7.9	9.8	11.7	14.3	15.9	18.3	20.1	21.7

Find the least squares estimate of the linear relation between x and y.

By Hooke's Law, the linear relation between x and y has the form $y = mx$.

x_i	y_i	x_iy_i	x_i^2
1.0	4.1	4.10	1.00
1.5	6.2	9.30	2.25
2.0	7.9	15.80	4.00
2.5	9.8	24.50	6.25
3.0	11.7	35.10	9.00
3.5	14.3	50.05	12.25
4.0	15.9	63.60	16.00
4.5	18.3	82.35	20.25
5.0	20.1	100.50	25.00
5.5	21.7	119.35	30.25
		504.65	126.25

$$\hat{m} = \frac{\sum x_iy_i}{\sum x_i^2}$$

$$\hat{m} = \frac{504.65}{126.25} = 4.00 \quad \text{(2 d.p.)}$$

Least squares estimate of $y = mx$ is $y = 4.00x$.

Least squares estimate of a linear relationship of the form $y = \alpha + \beta x$

Let the linear relationship between x and y be
$$y = \alpha + \beta x,$$
where α and β are unknown constants.

Suppose that
$$y = a + bx$$
is an estimate of $y = \alpha + \beta x$. As before, the method of least squares tries to find values of a and b which will minimise S the sum of the squares of the discrepancies.
$$S = \sum(y_i - a - bx_i)^2 \tag{1}$$
For any particular value of b, the value of a which minimises S is given by $\frac{dS}{da} = 0$, where $\frac{dS}{da}$ is obtained by differentiating S w.r.t. a, treating b as a constant. (This technique is called *partial differentiation*.)
$$\frac{dS}{da} = \sum 2(y_i - a - bx_i)(-1)$$
This is zero when
$$\sum(-y_i + a + bx_i) = 0$$
$$-\sum y_i + \sum a + b\sum x_i = 0$$
$$-n\bar{y} + na + bn\bar{x} = 0$$
Divide by n and rearrange: $\qquad a + b\bar{x} = \bar{y} \tag{2}$

For any particular value of a, the value of b which minimises S is given by $\dfrac{dS}{db} = 0$, where $\dfrac{dS}{db}$ is obtained by differentiating S w.r.t. b, treating a as a constant.

$$\frac{dS}{db} = \sum 2(y_i - a - bx_i)(-x_i)$$

This is zero when
$$\sum(-x_iy_i + ax_i + bx_i^2) = 0$$
$$-\sum x_iy_i + a\sum x_i + b\sum x_i^2 = 0$$
$$-\sum x_iy_i + an\bar{x} + b\sum x_i^2 = 0$$
$$an\bar{x} + b\sum x_i^2 = \sum x_iy_i \qquad (3)$$

Multiply (2) by $n\bar{x}$:
$$an\bar{x} + bn\bar{x}^2 = n\bar{x}\bar{y} \qquad (4)$$

Subtract (4) from (3) to eliminate a.

$$b\left(\sum x_i^2 - n\bar{x}^2\right) = \sum x_iy_i - n\bar{x}\bar{y}$$

$$b = \frac{\sum x_iy_i - n\bar{x}\bar{y}}{\sum x_i^2 - n\bar{x}^2}$$

The corresponding value of a may be found by substituting this value for b in equation (2). These values of a and b minimise S, but the proof of this assertion is beyond the scope of this book.

The line $\quad y = a + bx$

where $\quad b = \dfrac{\sum x_iy_i - n\bar{x}\bar{y}}{\sum x_i^2 - n\bar{x}^2} \quad$ and $\quad a = \bar{y} - b\bar{x}$

is called *the least squares estimate of the line* $y = \alpha + \beta x$.

The value $\hat{y}_0$, given by
$$\hat{y}_0 = a + bx_0,$$
is called *the least squares estimate for* $y_0 = \alpha + \beta x_0$, *the true value of* y *when* $x = x_0$.

The above formula for b is convenient to use in practical applications. An alternative version, which is more useful theoretically, is given below.

$$b = \frac{\sum(x_i - \bar{x})(y_i - \bar{y})}{\sum(x_i - \bar{x})^2}.$$

Consider the numerator of this formula for b.

$$\sum(x_i - \bar{x})(y_i - \bar{y}) = \sum(x_iy_i - x_i\bar{y} - \bar{x}y_i + \bar{x}\bar{y})$$
$$= \sum x_iy_i - \bar{y}\sum x_i - \bar{x}\sum y_i + \sum \bar{x}\bar{y}$$
$$= \sum x_iy_i - \bar{y}n\bar{x} - \bar{x}n\bar{y} + n\bar{x}\bar{y}$$
$$= \sum x_iy_i - n\bar{x}\bar{y}.$$

The denominator of the second formula for b may be written in the following form (see page 24, Book 1).

$$\sum (x_i - \bar{x})^2 = \sum x_i^2 - n\bar{x}^2.$$

Thus the two versions of the formula for b are equivalent.

Example 2

The resistance y ohms of a certain length of wire is linearly related to its temperature $x°C$. The resistance of the wire was measured at five different temperatures and the following results were obtained:

x	10	20	30	40	50
y	3.13	3.19	3.28	3.46	3.42.

a Find the least squares estimate of the linear relation between x and y.
b Find the least squares estimate of the true value of y when $x = 20$.

a

x_i	y_i	$x_i y_i$	x_i^2
10	3.13	31.3	100
20	3.19	63.8	400
30	3.28	98.4	900
40	3.46	138.4	1600
50	3.42	171.0	2500
150	16.48	502.9	5500

$$n = 5 \quad \bar{x} = \frac{150}{5} = 30 \quad \bar{y} = \frac{16.48}{5} = 3.296$$

$$b = \frac{\sum x_i y_i - n\bar{x}\bar{y}}{\sum x_i^2 - n\bar{x}^2}$$

$$b = \frac{502.9 - 5 \times 30 \times 3.296}{5500 - 5 \times 30^2}$$

$$b = \frac{8.5}{1000}$$

$$b = 0.0085$$

$$a = \bar{y} - b\bar{x}$$

$$a = 3.296 - 0.0085 \times 30$$

$$a = 3.041$$

Least squares estimate for the linear relationship between x and y is
$$y = 3.041 + 0.0085x.$$

147

b The least squares estimate for the true value of y when $x = 20$ is given by

$$\hat{y}_0 = a + bx_0$$
$$\hat{y}_0 = 3.041 + 0.0085 \times 20$$
$$\hat{y}_0 = 3.211.$$

Exercise 9.1

1 Two quantities x and y are such that $y = mx$. Values for y were measured at five controlled values of x.

x	1	2	3	4	5
y	1.8	5.4	7.2	10.5	11.6

a Find the least squares estimate of m.
b Find the least squares estimate of the true value of y when $x = 2.5$.

2 Two quantities x and y are such that $y = mx$. Four values of y were measured for pre-determined values of x.

x	0.5	1.0	1.5	2.0
y	1.2	1.5	2.7	3.3

Find the least squares estimates for m and the true value of y when $x = 1.5$.

3 The following table gives experimentally observed values of two variables x and y which are known to be linearly related. It may be assumed that the values of x are accurate and that the values of y may be subject to error.

x	10	15	20	25	30	35	40
y	4	7	12	17	21	24	27

Find the least squares estimates for
a the linear relation between x and y,
b the true value of y when $x = 20$.

4 Two quantities x and y are linearly related. Twenty-two pairs of values of x and y were experimentally obtained and the following results calculated:

$$\sum x = 176 \qquad \sum y = 396 \qquad \sum x^2 = 1429 \qquad \sum xy = 3210.$$

Assuming that the values of x are accurate and that the values of y may be in error, find the least squares estimates of
a the linear relation between x and y,
b the true value of y when $x = 10$.

5 Given the following five pairs of values of x and y

x	0	1	2	3	4
y	1	2	5	10	17

find the least squares estimate of the linear relation between x and y. Explain why your answer is absurd.

9.2 Least squares estimators

In this section it will be assumed that x and y are two algebraic (i.e. non-random) variables which are linearly related in the form $y = \alpha + \beta x$.

Experimentally obtained pairs of values, (x_1, y_1), (x_2, y_2), ..., (x_n, y_n), are such that the values $x_1, x_2, ..., x_n$ are accurate and the values $y_1, y_2, ..., y_n$ are subject to measurement errors $e_1, e_2, ..., e_n$. Furthermore, it will be assumed that these measurement errors are independent observations of a random variable ε which is normally distributed with mean zero and variance σ^2. Under these assumptions the values $y_1, y_2, ..., y_n$ may be regarded as observations of independent random variables $Y_1, Y_2, ..., Y_n$ such that

$$Y_i = \alpha + \beta x_i + \varepsilon, \qquad i = 1, 2, ..., n.$$

Since $\varepsilon \sim N(0, \sigma^2)$ and α, β, x_i are constant

$$Y_i \sim N(\alpha + \beta x_i, \sigma^2)$$

where $\alpha + \beta x_i$ is the true value of y when $x = x_i$.

The random variables given by

$$A = \bar{Y} - B\bar{x}$$

$$B = \frac{\sum(x_i - \bar{x})(Y_i - \bar{Y})}{\sum(x_i - \bar{x})^2}$$

$$\hat{Y}_0 = A + Bx_0$$

are the *least squares estimators for* α, β, and y_0, the true value of y when $x = x_0$, respectively. In the previous section the formulae for the corresponding estimates were established. It should be noted that the capital letters A, B, $\hat{Y}_0$, Y_i, $\bar{Y}$ are used to denote the random variables in the above formulae and the lower case letters $\bar{x}$, x_i, x_0 are used to denote the non-random variables.

The sampling distributions of A, B, $\hat{Y}_0$ will be derived in the following sections. The derivations are difficult and may be omitted on a first reading, but it is important to understand the assumptions which underlie the derivations.

Before proceeding, the reader is reminded of the following results which will be frequently used in the derivations:

1 $E[aY + b] = aE[Y] + b$

2 $V[aY + b] = a^2V[Y]$

Also the distribution of a linear combination of independent normally distributed random variables is itself normally distributed such that

3 $E\left[\sum a_i Y_i\right] = \sum a_i E[Y_i]$ (independence not relevant)

4 $V\left[\sum a_i Y_i\right] = \sum a_i^2 V[Y_i]$ (independence relevant)

A particular application of **3** is used to find $E[\bar{Y}]$ as follows:

$$E[\bar{Y}] = E\left[\frac{\sum Y_i}{n}\right]$$

$$= \sum \frac{1}{n} E[Y_i] \quad \text{(using 3)}$$

$$= \frac{1}{n} \sum (\alpha + \beta x_i)$$

$$= \frac{1}{n}\left(n\alpha + \beta \sum x_i\right).$$

Since $\dfrac{\sum x_i}{n} = \bar{x}$, it follows that

5 $E[\bar{Y}] = \alpha + \beta \bar{x}.$

Sampling distribution of B

$$E[B] = E\left[\frac{\sum (x_i - \bar{x})(Y_i - \bar{Y})}{\sum (x_i - \bar{x})^2}\right]$$

$$= \frac{1}{\sum (x_i - \bar{x})^2} E\left[\sum (x_i - \bar{x})(Y_i - \bar{Y})\right] \quad \text{(using 1)}$$

$$= \frac{1}{\sum (x_i - \bar{x})^2} \sum (x_i - \bar{x}) E[Y_i - \bar{Y}] \quad \text{(using 3)}$$

$$= \frac{1}{\sum (x_i - \bar{x})^2} \sum (x_i - \bar{x})(E[Y_i] - E[\bar{Y}])$$

$$= \frac{1}{\sum (x_i - \bar{x})^2} \sum (x_i - \bar{x})(\alpha + \beta x_i - \alpha - \beta \bar{x}) \quad \text{(using 5)}$$

$$= \frac{1}{\sum (x_i - \bar{x})^2} \sum (x_i - \bar{x})\beta(x_i - \bar{x})$$

$$= \frac{1}{\sum (x_i - \bar{x})^2} \cdot \beta \sum (x_i - \bar{x})^2$$

$$= \beta$$

This establishes that B is an unbiased estimator for β.

The above formula for B is not in a very convenient form from which to derive its variance. So, before proceeding with the derivation of $V[B]$, an alternative expression for B will be established.

Consider $\sum(x_i - \bar{x})(Y_i - \bar{Y}) = \sum(x_i - \bar{x})Y_i - \sum(x_i - \bar{x})\bar{Y}$

$$= \sum(x_i - \bar{x})Y_i - \bar{Y}\sum(x_i - \bar{\cdot}).$$

Since $\sum(x_i - \bar{x}) = \sum x_i - \sum \bar{x} = n\bar{x} - n\bar{x} = 0$

$$\sum(x_i - \bar{x})(Y_i - \bar{Y}) = \sum(x_i - \bar{x})Y_i.$$

Therefore $\quad B = \dfrac{\sum(x_i - \bar{x})Y_i}{\sum(x_i - \bar{x})^2}$

is an alternative formula for B.

$$V[B] = V\left[\frac{\sum(x_i - \bar{x})Y_i}{\sum(x_i - \bar{x})^2}\right]$$

$$= \frac{1}{\left\{\sum(x_i - \bar{x})^2\right\}^2}V\left[\sum(x_i - \bar{x})Y_i\right] \qquad \text{(using 2)}$$

$$= \frac{1}{\left\{\sum(x_i - \bar{x})^2\right\}^2}\sum(x_i - \bar{x})^2 V[Y_i] \qquad \text{(using 4)}$$

$$= \frac{1}{\left\{\sum(x_i - \bar{x})^2\right\}^2}\sum(x_i - \bar{x})^2 \sigma^2$$

$$= \frac{1}{\left\{\sum(x_i - \bar{x})^2\right\}^2}\sigma^2\sum(x_i - \bar{x})^2$$

$$= \frac{\sigma^2}{\sum(x_i - \bar{x})^2}$$

Since B is a linear combination of the n independent normally distributed random variables Y_i $(i = 1, 2, \ldots, n)$, it is itself normally distributed.

Therefore the sampling distribution of B is given by

$$B \sim N\left(\beta, \frac{\sigma^2}{\sum(x_i - \bar{x})^2}\right).$$

Sampling distribution of A

A is a linear combination of independent normally distributed random variables so it is itself normally distributed.

$$E[A] = E[\bar{Y} - B\bar{x}]$$
$$= E[\bar{Y}] - \bar{x}E[B] \qquad \text{(using 3)}$$
$$= (\alpha + \beta\bar{x}) - \bar{x}\beta \qquad \text{(using 5)}$$
$$= \alpha$$

Thus A is an unbiased estimator for α.

$$V[A] = V[\bar{Y} - B\bar{x}]$$

Since Y and B are not clearly independent some rearrangement is necessary.

$$V[A] = V\left[\frac{\sum Y_i}{n} - \bar{x}\frac{\sum(x_i - \bar{x})Y_i}{\sum(x_i - \bar{x})^2}\right]$$

$$= V\left[\sum Y_i\left\{\frac{1}{n} - \frac{\bar{x}(x_i - \bar{x})}{\sum(x_i - \bar{x})^2}\right\}\right]$$

$$= \sum\left\{\frac{1}{n} - \frac{\bar{x}(x_i - \bar{x})}{\sum(x - x)^2}\right\}^2 V[Y_i] \qquad \text{(using 4)}$$

$$= \sum\left\{\frac{1}{n^2} - \frac{2\bar{x}(x_i - \bar{x})}{n\sum(x_i - \bar{x})^2} + \frac{\bar{x}^2(x_i - \bar{x})^2}{\left[\sum(x_i - \bar{x})^2\right]^2}\right\}\sigma^2$$

$$= \sigma^2\left\{\sum\frac{1}{n^2} - \frac{2\bar{x}\sum(x_i - \bar{x})}{n\sum(x_i - \bar{x})^2} + \frac{\bar{x}^2\sum(x_i - \bar{x})^2}{\left[\sum(x_i - \bar{x})^2\right]^2}\right\}$$

Since $\sum(x_i - \bar{x}) = 0$

$$V[A] = \sigma^2\left\{n\cdot\frac{1}{n^2} + \frac{\bar{x}^2}{\sum(x_i - \bar{x})^2}\right\}.$$

An alternative expression for $V[A]$ is derived below.

$$V[A] = \sigma^2\left\{\frac{\sum(x_i - \bar{x})^2 + n\bar{x}^2}{n\sum(x_i - \bar{x})^2}\right\}$$

$$= \sigma^2\left\{\frac{\sum x_i^2 - n\bar{x}^2 + n\bar{x}^2}{n\sum(x_i - \bar{x})^2}\right\}$$

$$= \left\{\frac{\sigma^2\sum x_i^2}{n\sum(x_i - \bar{x})^2}\right\}$$

Thus the sampling distribution of A is given by

$$A \sim N\left(\alpha, \frac{\sigma^2 \sum x_i^2}{n \sum (x_i - \bar{x})^2}\right)$$

or

$$A \sim N\left(\alpha, \sigma^2 \left\{\frac{1}{n} + \frac{\bar{x}^2}{\sum (x_i - \bar{x})^2}\right\}\right).$$

Sampling distribution of $\hat{Y}_0$

$\hat{Y}_0$ is a linear combination of independent normally distributed random variables and is itself normally distributed.

$$E[\hat{Y}_0] = E[A + Bx_0]$$
$$= E[A] + x_0 E[B] \qquad \text{(using 3)}$$
$$= \alpha + \beta x_0$$

Since $y_0 = \alpha + \beta x_0$ is the true value of y when $x = x_0$, $\hat{Y}_0$ is an unbiased estimator for y_0.

$$V[\hat{Y}_0] = V[A + Bx_0]$$

A and B are not independent so some rearrangement is necessary.

$$V[\hat{Y}_0] = V[\bar{Y} - B\bar{x} + Bx_0]$$
$$= V[\bar{Y} + (x_0 - \bar{x})B]$$
$$= V\left[\frac{\sum Y_i}{n} + (x_0 - \bar{x})\frac{\sum (x_i - \bar{x})Y_i}{\sum (x_i - \bar{x})^2}\right]$$
$$= V\left[\sum Y_i\left\{\frac{1}{n} + \frac{(x_0 - \bar{x})(x_i - \bar{x})}{\sum (x_i - \bar{x})^2}\right\}\right] \qquad \text{(using 4)}$$
$$= \sum \left\{\frac{1}{n} + \frac{(x_0 - \bar{x})(x_i - \bar{x})}{\sum (x_i - \bar{x})^2}\right\}^2 V[Y_i]$$
$$= \sum \left\{\frac{1}{n^2} + \frac{2}{n}\frac{(x_0 - \bar{x})(x_i - \bar{x})}{\sum (x_i - \bar{x})^2} + \frac{(x_0 - \bar{x})^2(x_i - \bar{x})^2}{\left[\sum (x_i - \bar{x})^2\right]^2}\right\}\sigma^2$$
$$= \sigma^2 \left\{\sum \frac{1}{n^2} + \frac{2}{n}\frac{(x_0 - \bar{x})\sum (x_i - \bar{x})}{\sum (x_i - \bar{x})^2} + \frac{(x_0 - \bar{x})^2 \sum (x_i - \bar{x})^2}{\left[\sum (x_i - \bar{x})^2\right]^2}\right\}$$

Since $\sum (x_i - \bar{x}) = 0$

$$V[\hat{Y}_0] = \sigma^2 \left\{n \cdot \frac{1}{n^2} + \frac{(x_0 - \bar{x})^2}{\sum (x_i - \bar{x})^2}\right\}$$

An alternative expression for $V[\hat{Y}_0]$ is given below.

$$V[\hat{Y}_0] = \sigma^2 \left\{ \frac{\sum (x_i - \bar{x})^2 + n(x_0 - \bar{x})^2}{n \sum (x_i - \bar{x})^2} \right\}$$

The sampling distribution of $\hat{Y}_0$ is given by

$$\hat{Y}_0 \sim N\left(y_0, \sigma^2 \left\{ \frac{\sum (x_i - \bar{x})^2 + n(x_0 - \bar{x})^2}{n \sum (x_i - \bar{x})^2} \right\} \right)$$

or $\qquad \hat{Y}_0 \sim N\left(y_0, \sigma^2 \left\{ \frac{1}{n} + \frac{(x_0 - \bar{x})^2}{\sum (x_i - \bar{x})^2} \right\} \right).$

Confidence intervals for α, β and y_0

Symmetric 95% confidence intervals for α, β and y_0 are given by

For α: $a \pm 1.96 \times SE[A]$

For β: $b \pm 1.96 \times SE[B]$

For y_0: $\hat{y}_0 \pm 1.96 \times SE[\hat{Y}_0]$

Other symmetric confidence intervals may be obtained by substituting the appropriate numbers for 1.96.

Unknown σ^2

When σ^2 is unknown and is replaced in the formulae by an estimate based on the experimental values, the distributions of A, B and $\hat{Y}_0$ are no longer normal. This case is beyond the scope of this book but it may be noted that the statistics will have t-distributions.

Example 3

The linear relation between the length y m of an elastic string and its tension x N is given by $y = \alpha + \beta x$, where α and β are unknown constants. In an experiment to estimate α and β, ten pairs of values for (x, y) were obtained and the following results were calculated:

$$\sum x = 30, \qquad \sum y = 15.00, \qquad \sum x^2 = 110, \qquad \sum xy = 46.20.$$

The x measurements are accurate and the y measurements are subject to independent random errors, each normally distributed with mean zero and standard deviation 0.01 m.

a   Find the least squares estimate of $y = \alpha + \beta x$.

b   Calculate a 95% confidence interval for β.

c   Deduce a 95% confidence interval for the increase in the true value of y when x is increased from 1 to 5.

d   Determine an unbiased estimate for y_0 the true value of y when $x = 4$.

e   Test, using a 5% significance level, the null hypothesis H_0: $y_0 = 1.55$ against the alternative hypothesis H_1: $y_0 \neq 1.55$.

a $\bar{x} = \dfrac{30}{10} = 3$ $\qquad\qquad \bar{y} = \dfrac{15.00}{10} = 1.5$

$b = \dfrac{\sum xy - n\bar{x}\bar{y}}{\sum x^2 - n\bar{x}^2}$

$b = \dfrac{46.20 - 10 \times 3 \times 1.5}{110 - 10 \times 3^2}$

$b = \dfrac{1.2}{20}$

$b = 0.06$

$a = \bar{y} - b\bar{x}$

$a = 1.5 - 0.06 \times 3$

$a = 1.32$

The least squares estimate of $y = \alpha + \beta x$: $\qquad y = 1.32 + 0.06x$

b $B \sim N\left(\beta, \dfrac{\sigma^2}{\sum(x_i - \bar{x})^2}\right)$

$\sum(x_i - \bar{x})^2 = 20$, evaluated above as the denominator of the expression for b

$\quad V[B] = \dfrac{0.01^2}{20}$

$\quad SE[B] = 0.00223606$

95% confidence limits for β are given by

$\qquad b \pm 1.96 \times SE[B]$

$\qquad 0.06 \pm 1.96 \times 0.00223606$

$\qquad 0.06 \pm 0.00438269$

95% confidence interval for β: $\qquad (0.0556, 0.0644)$

c When $x = 1$ the true value of $y = \alpha + \beta$.

When $x = 5$ the true value of $y = \alpha + 5\beta$.

The increase in the true value of y when x is increased from 1 to 5 is $(\alpha + 5\beta) - (\alpha + \beta) = 4\beta$.

The 95% confidence limits for 4β are obtained by multiplying the 95% confidence limits for β by 4, therefore the required confidence limits are given by $4(0.06 \pm 0.00438269)$.

Required 95% confidence interval: $\qquad (0.222, 0.258)$.

d Unbiased estimate for y_0 is given by $\hat{y}_0 = a + bx_0$, where $x_0 = 4$.

$\qquad \hat{y}_0 = 1.32 + 0.06 \times 4$

$\qquad \hat{y}_0 = 1.56$

e $V[\hat{Y}_0] = \sigma^2 \left[\dfrac{\sum(x_i - \bar{x})^2 + n(x_0 - \bar{x})^2}{n\sum(x_i - \bar{x})^2} \right]$

$V[\hat{Y}_0] = 0.01^2 \left[\dfrac{20 + 10 \times (4-3)^2}{10 \times 20} \right]$

$V[\hat{Y}_0] = 1.5 \times 10^{-5}$

$SE[\hat{Y}_0] = 0.00387298$

$Z = \dfrac{\hat{Y}_0 - y_0}{SE[\hat{Y}_0]} \sim N(0,1)$

$H_0: y_0 = 1.55$ $\qquad\qquad$ $H_1: y_0 \neq 1.55$

$z = \dfrac{1.56 - 1.55}{0.00387289}$

$z = 2.58 > 1.96$, the result is significant at the 5% level.

Conclude that $y_0 > 1.55$.

Exercise 9.2

1 The linear relation between x and y is given by $y = \alpha + \beta x$, where α, β are unknown constants. In an experiment to estimate α and β twenty pairs of values (x, y) were obtained and the following results calculated:
$$\sum x = 110, \qquad \sum y = 300, \qquad \sum x^2 = 770, \qquad \sum xy = 2046.$$

Given that the x measurements are accurate and that the y measurements are subject to independent random errors, each normally distributed with mean zero and standard deviation 0.1, determine

a the least squares estimate of $y = \alpha + \beta x$,

b a 95% confidence interval for β,

c a 90% confidence interval for α,

d a 99% confidence interval for the true value of y when $x = 7.5$.

2 The linear relation between x and y is given by $y = \alpha + \beta x$. Fifteen pairs of values of x and y were obtained experimentally and the following results calculated:
$$\sum x = 120, \qquad \sum y = 138, \qquad \sum x^2 = 1240, \qquad \sum xy = 880.$$

Given that the x measurements are accurate and that the y measurements are subject to independent random errors, each normally distributed with mean zero and standard deviation 0.5, determine

a the least squares estimate of the line $y = \alpha + \beta x$,

b a 90% confidence interval for β,

c a 95% confidence interval for α,

d a 95% confidence interval for the true value of y when $x = 10$.

3 Two variables x and y are linearly related. Ten pairs of values (x, y) were experimentally obtained and the following results calculated:

$$\sum x = 500, \qquad \sum y = 685, \qquad \sum x^2 = 27000, \qquad \sum xy = 36650.$$

The measurements of x are accurate and the measurements of y are subject to independent random errors, each normally distributed with mean zero and standard deviation 0.2.

a Find the least squares estimate of y_0, the true value of y when $x = 45$.

b Test, using a 5% significance level, the null hypothesis H_0: $y_0 = 63$ against the alternative hypothesis H_1: $y_0 \neq 63$.

c Calculate a 95% confidence interval for the true value of the gradient of the line.

d Deduce a 95% confidence interval for the increase in the true value of y as x increases from 40 to 60.

4 At time t the velocity v of a body moving in a straight line with uniform acceleration f is given by the formula $v = u + ft$, where u is a constant. In an experiment, the velocity v was measured at certain pre-determined times t and the results are shown in the following table.

t	0	1	2	3	4
v	3.2	7.1	10.9	15.1	18.7

a Find the least squares estimates for u and f.

b Given that the measurements of t are accurate and that the measurements of v are subject to independent random errors, each normally distributed with mean zero and standard deviation 0.2, determine whether there is sufficient evidence at the 5% level to support the hypothesis that the value of $f < 4$.

5 The linear relation between x and y is of the form $y = mx$, where m is an unknown constant. In an experiment n pairs of values (x, y) were obtained. Assuming that the measurements x are accurate and that the measurements y are subject to independent random errors, each normally distributed with mean zero and standard deviation σ^2, show that the least squares estimator $\hat{M}$ of m, where

$$\hat{M} = \frac{\sum x_i Y_i}{\sum x_i^2},$$

is unbiased and that $V[\hat{M}] = \dfrac{\sigma^2}{\sum x_i^2}.$

The results of the experiment are shown below.

x_i	2	4	6	8
y_i	3.1	6.5	9.5	13.0

a Given that $\sigma = 0.1$, calculate a 95% confidence interval for m.

b Deduce a 95% confidence interval for the true value of y when $x = 5$.

9.3 Non-linear relations

Many relations between two variables are non-linear in form. Nevertheless it may be possible to transform one or both variables in such a way that there is a linear relation between the transformed variables and then the methods described above may be used to estimate the constants in the non-linear relationship. A number of examples will provide clarification.

The equation $pv = k$, which arises from Boyle's Law, may be written in the form $p = k \cdot \dfrac{1}{v}$ and there is a linear relation between p and the transformed variable $\dfrac{1}{v}$.

The more general formula for the relation between the pressure p and the volume v of a given mass of gas at constant temperature is $pv^\gamma = k$. By taking logs of both sides, this may be written in the form $\log p + \gamma \log v = \log k$. This is a linear relation between the transformed variables $\log p$ and $\log v$.

The equation $s = \alpha v + \beta v^2$ may be written in the form $\dfrac{s}{v} = \alpha + \beta v$, which is a linear relation between v and the transformed variable $\dfrac{s}{v}$.

Example 4

The relation between x and y is known to be of the form
$$y = ke^{\beta x}$$
where k and β are unknown constants. In an experiment to estimate k and β, values of y were measured at five controlled values of x.

x	1	2	3	4	5
y	6.2	7.4	8.9	11.2	13.6

Transform the above relation into linear form and use the method of least squares to find estimates for k and β, giving your answers to two decimal places.

Taking logs (to the base e) of both sides of the equation $y = ke^{\beta x}$.
$$\log y = \log k + \beta x$$

Writing $w = \log y$ and $\alpha = \log k$, a linear relation between w and x is obtained.
$$w = \alpha + \beta x$$

x	y	w	x^2	xw
1	6.2	1.8245493	1	1.8245493
2	7.4	2.0014800	4	4.0029600
3	8.9	2.1860513	9	6.5581538
4	11.2	2.4159138	16	9.6636551
5	13.6	2.6100698	25	13.0503490
15		11.0380642	55	35.0996672

$$\bar{x} = \frac{15}{5} = 3 \qquad \bar{w} = \frac{11.0380642}{5} = 2.2076128$$

$$b = \frac{\sum xw - n\bar{x}\bar{w}}{\sum x^2 - n\bar{x}^2}$$

$$b = \frac{35.0996672 - 5 \times 3 \times 2.2076128}{55 - 5 \times 3^2}$$

$$b = \frac{1.985475}{10}$$

$$b = 0.1985475$$

$$a = \bar{w} - b\bar{x}$$

$$a = 2.2076128 - 0.1985475 \times 3$$

$$a = 1.6119703$$

Since $\alpha = \log k$ $\qquad k = e^{\alpha}$.

Therefore $\hat{k}$, the estimate for k, is given by $\hat{k} = e^{1.6119703} = 5.012678$.

The estimates of k and β are given by

$$\hat{k} = 5.01 \qquad b = 0.20 \qquad (2\,\text{d.p.})$$

respectively.

Exercise 9.3

1 The relation between x and y is known to be of the form
$$y = ke^{\beta x},$$
where k and β are unknown constants. Values of y were measured at four controlled values of x.

x	1.0	1.5	2.0	2.5
y	3.7	2.2	1.4	0.8

Transform the above relation into linear form and use the method of least squares to find estimates for k and β.

2 The relation between x and y is known to be of the form
$$xy = \alpha x + \beta,$$
where α and β are unknown constants. Values of y were measured at five controlled values of x.

x	10	20	25	40	50
y	18.4	12.7	11.8	10.1	9.9

Transform the above relation into linear form and find the least squares estimates for α and β.

3 The stopping distance, s m, of a car driven by a particular person is related to its speed v km/h by the formula

$$s = \alpha v + \beta v^2,$$

where α and β are unknown constants. Values of s were measured at five controlled values of v.

v	20	40	50	80	100
s	11.7	26.4	34.2	59.6	81.2

Transform the above relation into linear form and use the method of least squares to find an estimate of the stopping distance when the speed is 30 km/h.

4 In order to estimate the values of the unknown constants γ and k in the formula

$$pv^\gamma = k,$$

values of p were measured at four controlled values of v.

v	10	15	20	25
p	75	33	19	12

Transform the above relation into linear form and use the method of least squares to find estimates for γ and k.

5 In order to estimate the value of the unknown constant k in the formula

$$pv = k,$$

values of p were measured at four controlled values of v.

v	10	15	20	25
p	76	51	40	32

Transform the above relation into linear form and use the method of least squares to find an estimate for k.

9.4 Linear regression

In this section a type of problem closely related to the previous work in this chapter will be considered. Suppose that Y is a random variable whose distribution depends upon the value of an algebraic (i.e. non-random) variable x. Some examples of such pairs of variables are listed below.

1 Y is the yield in tonnes from a wheat field of specified size when treated with x tonnes of fertiliser.

2 Y is the duration of relief from pain experienced by a patient when given a dose of x units of a pain-killing drug.

3 Y is the profit made from the sale of a product when £x is spent on advertising the product.

In each case, the value of x, which may be controlled or measured accurately, influences but does not fully determine the value of Y; the random component of Y is an inherent feature of the process under observation. It may be noted that the work in the previous sections of this chapter may be considered to be a special case in which the random component of Y is entirely due to measurement error.

Consideration will be restricted to the case when the relation between Y and x takes the form

$$Y = \alpha + \beta x + \varepsilon,$$

where α, β are unknown constants and ε is a random term having zero mean. In this case, the mean of Y is a linear function of x given by

$$E[Y] = \alpha + \beta x.$$

This line is called *the regression line of Y on x*. Its gradient β is sometimes called the *regression coefficient*.

The diagram below shows this regression line together with sketches of the probability density functions of Y for three values x_1, x_2, x_3. Also marked with crosses are the points (x_1, y_1), (x_2, y_2) and (x_3, y_3), where y_1, y_2, y_3 are three observed values of Y corresponding to x_1, x_2, x_3, respectively.

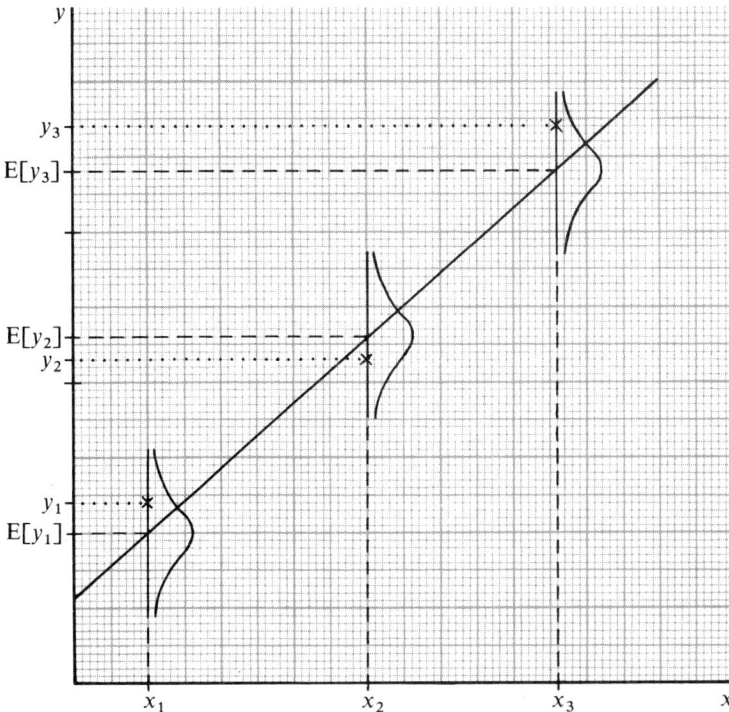

STATISTICS 2

Using the principle of least squares and the same notation as in the previous sections of this chapter, the least squares estimate for the regression line is

$$y = a + bx,$$

where a and b are given by same formulae as stated in **9.1**. This line is called *the sample least squares regression line of y on x.* Making the further assumption that for each value of x the distribution of Y is normal with known variance σ^2, which is independent of the value of x, then the least squares estimators A, B, $\hat{Y}_0$ have the same sampling distributions as those derived in **9.2**. There is one important distinction to be made: in this section the least squares estimate $\hat{y}_0 = a + bx_0$ now provides an estimate for the mean value of the random variable Y when $x = x_0$, whereas in **9.2**, $\hat{y}_0$ provided an estimate for the true value of the algebraic variable y when $x = x_0$.

Example 5

The following table gives the observed value y min of the duration of pain relief when x units of a pain-killing drug were administered to each of 12 patients.

x	1	2	3	4
y	7.2	19.1	30.5	41.3
	6.9	18.3	30.8	42.5
	7.1	18.7	31.4	43.2

a Find the sample least squares regression line of y on x.

Given that the duration of pain-relief for each value of x is normally distributed with standard deviation 0.5 min, find 90% confidence intervals for the mean duration of pain-relief when 3 units of drug are given,

b using all the data in the table,

c using only the three observed values of y when $x = 3$.

State, with a reason which of these two confidence intervals is preferable.

a Using a calculator the following summations are calculated:

$$n = 12 \qquad \sum x = 30 \qquad \bar{x} = 2.5 \qquad \sum x^2 = 90$$

$$\sum y = 297 \qquad \bar{y} = 24.75 \qquad \sum xy = 919.5$$

$$b = \frac{\sum xy - n\bar{x}\bar{y}}{\sum x^2 - n\bar{x}^2}$$

$$b = \frac{919.5 - 12 \times 2.5 \times 24.75}{90 - 12 \times 2.5^2}$$

$$b = \frac{177}{15}$$

$$b = 11.8$$

$$a = \bar{y} - b\bar{x}$$
$$a = 24.75 - 11.8 \times 2.5$$
$$a = -4.75$$

The least squares regression line of y on x: $y = -4.75 + 11.8x$.

b Using all the data in the table.

When $x = 3$ $\qquad \hat{y}_0 = -4.75 + 11.8 \times 3$

$$\hat{y}_0 = 30.65$$

$$V[\hat{Y}_0] = \sigma^2 \left[\frac{\sum(x_i - \bar{x})^2 + n(x_0 - \bar{x})^2}{n\sum(x_i - \bar{x})^2} \right]$$

$$= 0.5^2 \left[\frac{15 + 12 \times (3 - 2.5)^2}{12 \times 15} \right]$$

$$= 0.025$$

$$SE[\hat{Y}_0] = 0.1581138$$

90% confidence limits for $E[Y]$ when $x = 3$ are given by

$$\hat{y}_0 \pm 1.645 \times SE[\hat{Y}_0]$$
$$30.65 \pm 1.645 \times 0.1581138$$
$$30.65 \pm 0.26 \quad \text{(2 d.p.)}$$

90% confidence interval for $E[Y]$ when $x = 3$ is (30.39, 30.91)

c Using only the three observed values of y when $x - 3$.

$$\bar{y}_3 = \frac{(30.5 + 30.8 + 31.4)}{3}$$
$$\bar{y}_3 = 30.9$$

$$V[\bar{Y}_3] = \frac{\sigma^2}{n}$$

$$= \frac{0.5^2}{3}$$

$$= 0.08\dot{3}$$

$$SE[\bar{Y}_3] = 0.2886751$$

90% confidence limits are given by

$$30.9 \pm 1.645 \times 0.2886751$$
$$30.9 \pm 0.47 \quad \text{(2 d.p.)}$$

90% confidence interval for $E[Y]$ when $x = 3$ is (30.43, 31.37)

The interval in **b** is preferable to that in **c** because it has a smaller width.

STATISTICS 2

Linear regression when both variables are random

This case will not be considered in detail, but, clearly, there are many practical situations in which the relationship between two associated random variables is of interest. Some examples are given below.

1   X and Y are the height and weight of a randomly chosen man.
2   X and Y are the ages of the husband and wife of a randomly chosen couple.
3   X and Y are the running cost per patient and the number of beds in a randomly chosen hospital.

In each case the letters X and Y could have been interchanged without loss of meaning and so the situation is not the same as in the previous sections when x was an algebraic variable.

Suppose that in n independent trials of a random experiment $(x_1, y_1), (x_2, y_2)$ ..., (x_n, y_n) are the observed values of (X, Y), and that a scatter diagram indicates a linear relation between X and Y. Using the same notation as before, it is possible to find the least squares regression line of y on x in the form $y = a + bx$ and to use it to estimate the mean value of Y for a particular value of X. However, in this case, because the interchangeability of x and y, it is also possible to find the *least squares regression line of x on y* given by

$$x = a' + b'y,$$

where
$$b' = \frac{\sum (x_i - \bar{x})(y_i - \bar{y})}{\sum (y_i - \bar{y})^2}$$
and
$$a' = \bar{x} - b'\bar{y},$$

and to use it to estimate the mean value of X for a particular value of Y. In general, these two regression lines are distinct but they always have the point $(\bar{x}, \bar{y})$ in common.

It is only when the pairs of observed values (x_i, y_i) have been obtained at random that *both* these regression lines can be considered. When, for example, the values of X have been controlled and the values of Y measured, then only the least squares regression line of y on x should be used.

Exercise 9.4

In each of the following questions it may be assumed that a linear regression model is appropriate.

1   In an experiment to test the effectiveness of a filtration system in removing waste solids from suspension in a fluid, on twenty-one occasions the flow was fixed at a particular rate, x gallons per minute, and the resulting percentage, y, of solid removed was measured.

x	5	6	7	8	9	10	11
y	23.1	20.2	20.3	18.2	16.0	15.5	13.0
	22.3	21.4	19.8	18.8	17.4	15.1	12.8
	23.9	20.8	19.3	17.6	16.7	14.7	13.2

a   Find the least squares regression line of y on x.

164

Given that, for each rate of flow, the percentage of waste removed is normally distributed with standard deviation 0.4, find a 90% confidence interval for the mean percentage of waste removed when the rate of flow is 10 gallons per minute

b using all the data in the table,

c using only the three observed values of y when $x = 10$.

2 The manager of a factory wishes to investigate the effect of noise level on the time taken to complete a particular task. An experiment is conducted in which the time (y min) taken by a worker to complete the task was measured for pre-set noise levels (x dB).

x	40	50	60	70	80
y	27.2	29.1	33.6	35.8	40.3

a Find the least squares regression line of y on x.

b Estimate the mean completion time when the noise level is 100 dB and comment upon your answer.

3 The table shows the yield per plot, y units, of a particular crop when treated with x units of fertiliser for each of fifteen plots.

x	3	4	5	6	7
y	37	43	52	56	61
	38	41	48	57	60
	36	41	50	55	60

Assuming that the yield per plot for any given amount of fertiliser is normally distributed with standard deviation 0.5 units, find a 95% confidence interval for the mean yield per plot when 6 units of fertiliser is applied,

a using all the data in the table,

b using only the three observed values of y when $x = 6$.

c State which of these two confidence intervals is preferable.

4 Values of a random variable Y were observed for five values of x controlled at 2, 3, 4, 5, 6 and the least squares regression line of y on x was calculated to be

$$y = 7.4 + 3.6x.$$

For each value of x, the observed value of Y is normally distributed with standard deviation 0.1.

a Find the mean $\bar{y}$ of the five observed values of Y.

b Calculate a 95% confidence interval for the mean value of Y when $x = 3.5$

5 The height (x m) and the weight (y kg) of ten randomly selected men are recorded in the following table.

x	1.69	1.71	1.72	1.73	1.74	1.75	1.75	1.78	1.80	1.83
y	63.6	67.2	65.4	70.8	71.2	75.6	76.1	80.2	84.9	89.0

Find the equation of the regression line appropriate for estimating the mean height of a man of weight 70 kg.

Miscellaneous Exercise 9

1 Two quantities x and y are linearly related. Ten pairs of experimental observations (x, y) were made and the following results were calculated:

$$\sum x = 90, \qquad \sum y = 415, \qquad \sum x^2 = 3300, \qquad \sum xy = 14193.$$

Assuming that the observations x are accurate and that the observations y are subject to independent random errors, each normally distributed with mean zero and standard deviation 0.5, determine

(i) the least squares estimate for the equation connecting x and y,

(ii) the 99 per cent symmetric confidence interval for the value of y when $x = 0$. *(JMB)*

2 A chemist set up an experiment to determine how a variable y varied with an associated variable x. In the experiment, x was set at the five values 0, 1, 2, 3, 4, respectively, and the corresponding values of y were observed. The chemist noted that the values of y increased fairly steadily with the increasing values of x, and, on applying the method of least squares to the results, the chemist produced the equation

$$y = 5.8 + 2.3x.$$

(i) Find the value of $\bar{y}$, the mean of the five observed values of y.

Suppose that the experimentally observed values of y are subject to independent random errors that are normally distributed with mean zero and standard deviation 1.1. Assuming that the true relationship connecting y and x is linear, calculate

(ii) a 95% confidence interval for the true value of y when $x = 4$,

(iii) a 90% confidence interval for the difference between the true values of y corresponding to $x = 1$ and $x = 4$, respectively. *(WJEC)*

3 A response variable y is known to be linearly related to the value of a variable x. To estimate the precise form of this relationship the response y is measured for each of six specified values of x, the results being as shown in the following table:

x	10	20	30	40	50	60
y	19	27	26	32	35	41

Whereas specified values of x are known to be accurate, the observed values of y are subject to independent normally distributed errors with mean zero and standard deviation 1.5.

(i) Calculate the least squares estimate of the linear relationship connecting x and y.

(ii) Calculate 95 per cent confidence limits for the slope of the relationship. Hence, or otherwise, find 95 per cent confidence limits for the increase in the value of y when x is increased by 20 units. (WJEC)

4 It is known that the heart rate of a patient treated with a certain drug is approximately linearly related to the dose administered, the precise form of this linear relationship being dependent on the patient. The following table shows the heart rate, y, in beats per minute of a patient treated with x grains of the drug on six distinct occasions.

x	1	2	3	3	4	5
y	50	60	70	80	90	100

a Calculate the least squares estimate of the linear relationship between y and x for this patient.

b Assuming that the heart rate determinations are subject to an error standard deviation of 4 beats per minute find

(i) a 95 per cent confidence interval for the increase in the patient's heart rate corresponding to an increase of 1 grain in the dose of the drug.

(ii) a 90 per cent confidence interval for the heart rate of the patient if he is given a dose of 3.5 grains. (WJEC)

5 The mass y grams of a certain chemical substance which is dissolved in a certain mass of water at a temperature of $x°C$ is given by the relation

$$y = \alpha + \beta x,$$

where α and β are unknown constants. In an experiment to estimate α and β, the temperature was carefully controlled at 10 different values between $0°C$ and $100°C$, and y was measured at each temperature value. The following calculations were made from the 10 observed pairs of values (x, y):

$$\sum x = 400, \quad \sum y = 688, \quad \sum x^2 = 19\,200, \quad \sum xy = 38\,720.$$

(i) Determine the least squares estimate of the equation connecting x and y.

(ii) Estimate the mass of the chemical dissolved when the temperature is $60°C$.

(iii) Given that the measurements of y are subject to a normally distributed random error having mean zero and variance 0.4, calculate the 99 per cent symmetric confidence interval for the true mass of the chemical dissolved when the temperature is $60°C$. (JMB)

6 At time t the velocity v of a body moving in a straight line with constant acceleration f is given by the formula

$$v = u + ft,$$

where u is a constant. In an experiment, the velocity v was measured at certain chosen times t and the results obtained are shown in the following table.

t	1	2	3	4	5
v	6.6	12.3	16.9	21.8	27.4

(i) Calculate, showing your method, the least squares estimates of f and u.

(ii) Given that the measurements of t are accurate and that the measurements of v are subject to independent random errors, each normally distributed with zero mean and standard deviation 0.2, determine whether there is sufficient evidence at the 5 per cent significance level to support the hypothesis that the value of f is greater than 5. *(JMB)*

7 Two variables x and y of interest in an experiment are known to be linearly related, but the coefficients in this relationship are not known. A series of 15 experiments was conducted in which 3 determinations of y were made for each of 5 values of x. The x-values are accurate but the determinations of the y-values are subject to independent random errors that are normally distributed with mean zero and standard deviation 1.25. The observed values in the series of experiments are shown in the following table.

x	1	2	3	4	5
y	19	17	12	9	3
	18	16	11	10	3
	21	17	13	7	4

The following quantities were calculated from the above data:

$$\sum x = 45, \qquad \sum y = 180, \qquad \sum x^2 = 165, \qquad \sum xy = 420.$$

(i) Calculate the equation of the least squares estimate of the linear relationship between x and y.

(ii) Estimate the true value of y when $x = 4$. Determine the standard error of this estimate. Explain why this estimate of y when $x = 4$ is preferable to that obtained from averaging the three observed values of y when $x = 4$.

(iii) Calculate a 90 per cent confidence interval for the true value of y when $x = 4$. *(WJEC)*

8 It is known that the two variables u and y are related in the form

$$y = \alpha + \frac{\beta}{u}$$

for values of u in the interval $0 < u \leqslant 1$. An experiment was conducted in which the value of y was observed for each of four specific values of u. The values of u and the corresponding observed values of y are given in the following table.

u	0.1	0.2	0.5	1.0
y	4.5	9.6	14.2	15.1

(i) By introducing a new variable instead of u, show that the least-squares estimate of β is equal to -1.2, and find the least-squares estimate of α.

(ii) The errors in the observed values of y are independent and normally distributed with mean zero and standard deviation 0.7. Calculate 95% confidence limits for

    (a) the true value of y when $u = 0.4$,

    (b) the true increase in the value of y when the value of u is increased from 0.5 to 1.0. *(WJEC)*

9 Measurements of the heart rate of a person are subject to independent random errors that are normally distributed with mean zero and standard deviation 0.5 beats per minute. It is also known that when a person is given x grains of a certain drug, where $0 < x \leqslant 5$, the person's heart rate, y beats per minute, is such that $y = \alpha + \beta x$, where α and β may vary from one person to another. The following table shows the measured heart rates of a particular person after being given various doses of the drug.

Dosage of drug (x grains)	1	2	3	4	5
Heart rate (y beats per min.)	52	63	75	89	101

(i) Determine the least squares estimates of α and β for this person.

(ii) Calculate a 90% confidence interval for this person's heart rate when no drug is administered.

(iii) The same experiment was conducted independently on another person using the same dosages of the drug (as given in the above table). From the results of this experiment the least squares estimate of the value of β for this second person was found to be 10.9. Calculate 95% confidence limits for $\beta_1 - \beta_2$, where β_1 and β_2 are the respective values of β for the first person and the second person. *(WJEC)*

10 Two variables x and y are known to be related by the formula $y = \lambda x$, where λ is an unknown constant. To obtain information on λ, experiments were performed with x having the values $x_1, x_2, \ldots, x_n$, and the corresponding values of y were measured. Denoting the measured values of y by $y_1, y_2, \ldots, y_n$, respectively, show that the least squares estimate, m, of λ is given by

$$m = \frac{\sum x_i y_i}{\sum x_i^2}.$$

Assuming that $y_i = \lambda x_i + \varepsilon_i$, where $\varepsilon_1, \varepsilon_2, \ldots, \varepsilon_n$ are independent random errors each having mean zero and variance σ^2, show that m is an unbiased estimate of λ and that its sampling variance is

$$\frac{\sigma^2}{\sum x_i^2}.$$

The actual values obtained in the experiments are shown in the following table.

x_i	1	2	3	4	5
y_i	2.6	5.1	7.7	9.8	12.7

(i) Calculate the value of m.

(ii) Assuming that the sampling distribution of m is normal and that $\sigma = 0.2$, determine the limits of the 95% symmetrical confidence intervals for λ and for the true value of y when $x = 3$. (*JMB*)

Chapter 10

Revision

This chapter contains brief notes and reminders about important points followed by exercises.

10.1 Poisson distribution

A discrete random variable X, whose probability function is given by

$$P(X = r) = e^{-\alpha}\frac{\alpha^r}{r!}, \qquad r = 0, 1, 2, \ldots$$

where $\alpha > 0$, has a Poisson distribution with parameter α.

The probability generating function G of $Po(\alpha)$ is given by

$$G(t) = e^{\alpha(t-1)}$$

$$\mu = G'(1) \qquad \text{and} \qquad \sigma^2 = G''(1) + \mu - \mu^2.$$

The distribution $Po(\alpha)$ has mean α and variance α.

Recursive formula $\qquad \dfrac{p_r}{p_{r-1}} = \dfrac{\alpha}{r} \qquad$ where $p_r = P(X = r)$

Approximations to the binomial distribution

If $X \sim B(n, p)$, n is large and

1   n is greater than the larger of $\dfrac{16p}{(1-p)}$ and $\dfrac{16(1-p)}{p}$, then X is approximately $N(np, np(1-p))$. Continuity corrections must be used.

2   p is small ($p < 0.1$ gives reasonable accuracy) then X is approximately $Po(np)$. No continuity correction needed.

3   p is close to 1 ($p > 0.9$ gives reasonable accuracy) then X', where $X' = n - X$, is approximately $Po(n(1-p))$. No continuity correction required.

Normal approximation to Po(a)

If $X \sim \text{Po}(\alpha)$ and n is large ($n > 16$ gives reasonable accuracy) then X is approximately $\text{N}(\alpha, \alpha)$. Continuity corrections must be used.

Additive property of independent Poisson variables

If X and Y are independent random variables having Poisson distributions with means α and β then $(X + Y)$ has the Poisson distribution with mean $(\alpha + \beta)$. This result may be extended to three or more independent Poisson variables.

Exercise 10.1

1 **Calculators are not to be used in this question.**
   The random variable X has the binomial distribution $B(n, p)$. Use the tables provided and, when necessary, a distributional approximation, to calculate $P(X \geqslant np)$ to three decimal places in each of the cases when
   (i) $n = 20$, $p = 0.3$,
   (ii) $n = 20$, $p = 0.8$,
   (iii) $n = 10$, $p = 0.25$,
   (iv) $n = 200$, $p = 0.015$.   (WJEC)

2 In each of n independent trials of an experiment the probability of an event A occurring is 0.05.
   (i) When $n = 10$, determine the probability that A occurs exactly once, giving your answer to three decimal places.
   (ii) When $n = 200$, use a suitable approximate method to determine the probability that A occurs not more than 10 times.   (JMB)

3 The demand per week for videos in a particular shop has the Poisson distribution with mean 4.
   (i) Find, to four decimal places, the probabilities that in a week the demand for videos at this shop will be
        (a) exactly 2,   (b) 5 or fewer.
   (ii) Assuming that the demands in successive weeks are independent, use an appropriate distributional approximation to calculate, to three decimal places, the probability that the demand for videos at the shop during a year of 52 weeks will be at least 220.

   Videos are delivered to the shop at the beginning of each week only, and the policy is to accept as many as are required for the stock level to be 5.
   (iii) Show that the most probable number of videos that will be *sold* in a week is 5, and find, to two decimal places, the expected number of videos that will be *sold* in a week.   (WJEC)

4 A discrete random variable X has the Poisson distribution given by

$$P(X = r) = e^{-a}\frac{a^r}{r!}, \qquad r = 0, 1, 2, \dots .$$

Prove that the mean and the variance of X are each equal to a.

When a trainee typist types a document the number of mistakes made on any one page is a Poisson variable with mean 3, independently of the number of mistakes made on any other page. Use tables, or otherwise, to find, to three significant figures,

(i) the probability that the number of mistakes on the first page is less than two,

(ii) the probability that the number of mistakes on the first page is more than four.

Find expressions in terms of e for

(iii) the probability that the first mistake appears on the second page,

(iv) the probability that the first mistake appears on the second page and the second mistake appears on the third page.

Evaluate these expressions, giving your answers to four significant figures.

When the typist types a 48-page document the total number of mistakes made by the typist is a Poisson variable with mean 144. Use a suitable approximate method to find, to three decimal places, the probability that this total number of mistakes is greater than 130. (*JMB*)

5 Independently for each day, the number of times, X, that a machine needs to be adjusted in a day has a Poisson distribution with mean 0.5

(i) Using the tables provided, or otherwise, find the smallest integer r for which $P(X \geqslant r) \leqslant 0.02$.

(ii) Calculate, to three decimal places, the probability that during a period of five days, the machine needs to be adjusted once on one day and twice on another day, but does not need to be adjusted on any of the remaining three days.

(iii) Given that on a particular day the machine was adjusted at least once, calculate to three decimal places, the probability that it was adjusted at least twice on that day.

The cost, £Y, of adjusting the machine on a day when it is adjusted X times is given by $Y = 3(2^X - 1)$.

(iv) Using the tables provided, or otherwise, find to three decimal places, the probability that the cost of adjusting the machine on a randomly chosen day will exceed £4.

(v) Calculate, to the nearest penny, the mean daily cost of adjusting the machine. (*JMB*)

10.2 The distribution of a function of a random variable

If X is a continuous random variable with probability density function f, distribution function F and range space $\{x: a \leqslant x \leqslant b\}$ then by definition

$$F(x) = P(X \leqslant x), \qquad \text{for all } x,$$
$$F(x) = 0, \qquad x < a,$$
$$F(x) = \int_a^x f(t)dt, \qquad a \leqslant x \leqslant b,$$
$$F(x) = 1, \qquad x > b.$$

Since integration is the inverse operation to differentiation

$$f(x) = F'(x).$$

The general method of finding the probability density function g of a random variable Y, which is a function h of X, is outlined below.

1 Start with $G(y) = P(Y \leqslant y)$, where G is the distribution function of Y.

2 Substitute $Y = h(X)$.

3 Rearrange the inequation to make X the subject.

4 Write $G(y)$ in terms of F.

5 Either differentiate using the chain rule and substitute for f or substitute for F and then differentiate (n.b. $F' = f$ and $G' = g$). In both cases $g(y)$ is found.

When h is a 1-1 function the result of **4** is

either $\qquad G(y) = F(h^{-1}(y)) \qquad$ when h is an increasing function,

or $\qquad G(y) = 1 - F(h^{-1}(y)) \qquad$ when h is a decreasing function.

These results may be learnt to form the basis of a shorter method.

An even quicker method may be used when the function h is a 1-1 function. It involves the application of the following formula

$$g(y) = f(x)\left|\frac{dx}{dy}\right|,$$

where it is implicit that x is expressed in terms of y on the right-hand side.

Exercise 10.2

1 The probability density function f of a random variable X is given by

$$f(x) = \frac{1}{4}, \qquad 0 \leqslant x \leqslant 4,$$
$$f(x) = 0, \qquad \text{otherwise.}$$

Find the probability density function of Y, where $Y = X^2$.

2 The probability density function f of a random variable X is given by

$$f(x) = \frac{x}{32}, \qquad 6 \leqslant x \leqslant 10,$$

$$f(x) = 0, \qquad \text{otherwise.}$$

Find the probability density function of Y, where $Y = \dfrac{(14 - X)}{2}$.

3 Records of times taken by a vehicle to complete a particular journey of 80 km indicate that the time X taken per journey is a continuous random variable having a probability density function f given by

$$f(x) = \frac{2}{x^2}, \qquad 1 \leqslant x \leqslant 2,$$

$$f(x) = 0, \qquad \text{otherwise.}$$

Find the probability density function of the average speed per journey.

4 A continuous random variable X has probability density function defined by

$$f(x) = \frac{2x}{a^2}, \qquad 0 \leqslant x \leqslant a$$

$$f(x) = 0, \qquad \text{otherwise.}$$

a Find the variance of X.

b Find the cumulative distribution function of X.

c If $Y = 1 - \left(\dfrac{X}{a}\right)$, find

   (i) $P(Y > X)$,

   (ii) the probability density function of Y. $\hfill$ (*WJEC*)

5 A continuous random variable X is distributed with probability density function

$$f(x) = 3x^{-4}, \qquad x \geqslant 1$$

$$f(x) = 0, \qquad \text{otherwise.}$$

Find

  (i) the cumulative distribution function of X,

  (ii) the mean and the variance of X.

The random variable Y is defined by $Y = X^{-1}$. By first finding the cumulative distribution function of Y, or otherwise, determine the probability density function of Y. Verify that $E(X)E(Y)$ and $E(XY)$ are not equal and explain why they have different values in this particular example. $\hfill$ (*WJEC*)

10.3 Joint distributions of discrete random variables

Two discrete random variables

Suppose that X and Y are two discrete random variables with range spaces $\{x_i: i = 1, 2, \ldots, n\}$ and $\{y_j: j = 1, 2, \ldots, m\}$ respectively, and that p is the function such that

$$p(x_i, y_j) = P(X = x_i \cap Y = y_j) \qquad \text{for all } i, j;$$

$$p(x_i, y_j) \geqslant 0 \qquad \text{for all } i, j;$$

$$\sum_{i=1}^{n} \sum_{j=1}^{m} p(x_i, y_j) = 1$$

then p is called the joint probability function of X and Y. A table giving the values of $p(x_i, y_j)$ for all pairs of values (x_i, y_j) is called the joint distribution of X and Y.

To prove the independence of X and Y it is necessary to show that

$$P(X = x_i \cap Y = y_j) = P(X = x_i) \cdot P(Y = y_j)$$

for all pairs of values (x_i, y_j).

To prove that X and Y are not independent it is only necessary to show that

$$P(X = x_i \cap Y = y_j) \neq P(X = x_i) \cdot P(Y = y_j)$$

for one pair of values (x_i, y_j).

The expected value of a function $f(X, Y)$ is

$$E[f(X, Y)] = \sum_{i=1}^{n} \sum_{j=1}^{m} p_{ij} f(x_i, y_j), \quad \text{where } p_{ij} = P(X = x_i \cap Y = y_j).$$

The result

$$E[aX + bY] = aE[X] + bE[Y]$$

is true whether or not X and Y are independent.

The results

$$V[aX + bY] = a^2 V[X] + b^2 V[Y]$$

$$E[XY] = E[X] \cdot E[Y]$$

are true when X and Y are independent. Occasionally they are true when X and Y are not independent.

Extensions to three or more variables

$$E[a_1 X_1 + a_2 X_2 + \cdots + a_n X_n] = a_1 E[X_1] + a_2 E[X_2] + \cdots + a_n E[X_n]$$

When $X_1, X_2, \ldots, X_n$ are independent

$$V[a_1 X_1 + a_2 X_2 + \cdots + a_n X_n] = a_1^2 V[X_1] + a_2^2 V[X_2] + \cdots + a_n^2 V[X_n]$$

$$E[X_1 X_2 \ldots X_n] = E[X_1] E[X_2] \ldots E[X_n]$$

Exercise 10.3

1 X is the number of 'sixes' obtained when a fair cubical die is thrown 36 times. Y has a Poisson distribution with mean 2 and is independent of X.

Find the mean and the variance of

a $2X + 1$ **b** $3Y - 4$ **c** $3X - 5Y$.

2 The random variable X has the Poisson distribution with mean 2.5.
 (i) Show that $5P(X = 2) = 6P(X = 3)$.
 (ii) Using tables, or otherwise, find the value of $P(X \leqslant 3)$ correct to four decimal places.
 (iii) The random variable Y is independent of X and has the Poisson distribution with mean 6. Find the mean and the standard deviation of $3Y - 2X$. *(WJEC)*

3 Three fair coins are tossed simultaneously and each coin showing a head is removed. The remaining coins, when there are any, are then tossed again. Let X denote the number of heads showing on the first toss and let Y denote the number of heads showing on the second toss; Y is taken to be zero if there is no second toss.
 (i) Find the values of p, a, b, c and d in the following tabular representation of the joint probability distribution of X and Y.

		x			
		0	1	2	3
	0	p	$6p$	$12p$	ap
y	1	$3p$	$12p$	bp	0
	2	$3p$	cp	0	0
	3	dp	0	0	0

 (ii) Show that X and Y are not independent.
 (iii) Find the mean and the standard deviation of $Z = X + Y$, the total number of heads showing in the two tosses. *(WJEC)*

4 The following table gives the joint probability distribution of two discrete random variables X and Y. Thus, for example,
$$P(X = 0, Y = 1) = \alpha \quad \text{and} \quad P(X = 1, Y = 3) = \beta.$$
You are given that $0 < \alpha < 1$ and $0 < \beta < 1$.

		x		
		0	1	2
	1	α	β	α
y	2	β	0	β
	3	α	β	α

 (i) Show that $\alpha + \beta = \frac{1}{4}$.

(ii) Find the numerical values of E(X) and E(Y) and show that
$E(XY) = E(X)E(Y)$.

(iii) Determine whether or not X and Y are independent.

(iv) Find the variance of $Z = X + Y$ in terms of α. *(WJEC)*

5 A discrete random variable X takes the values 0 and 1 only with probabilities p and $1 - p$, respectively, where $0 < p < 1$. Find E(X) and Var(X).

A second discrete random variable Y also takes the values 0 and 1 only. Find the joint probability distribution of X and Y in each of the cases when

(i) $P(Y = 0) = p$ and X and Y are independent,

(ii) $P(Y = 1 | X = 1) = P(Y = 0 | X = 0) = p$,
$P(Y = 1 | X = 0) = P(Y = 0 | X = 1) = 1 - p$.

In both cases, show that $E(X + Y) = E(X) + E(Y)$, and investigate, for varying values of p strictly between 0 and 1, the equality, or otherwise, of $E(XY)$ and $E(X)E(Y)$. *(WJEC)*

10.4 Joint distributions of continuous random variables

The result
$$E[a_1X_1 + a_2X_2 + \cdots + a_nX_n] = a_1E[X_1] + a_2E[X_2] + \cdots + a_nE[X_n]$$
and the results, when $X_1, X_2, \ldots, X_n$ are independent,
$$V[a_1X_1 + a_2X_2 + \cdots + a_nX_n] = a_1^2V[X_1] + a_2^2V[X_2] + \cdots + a_n^2V[X_n]$$
$$E[X_1X_2\ldots X_n] = E[X_1]E[X_2]\ldots E[X_n]$$
are also applicable when some, or all, of $X_1, X_2, \ldots, X_n$ are continuous random variables.

Linear combination of independent normal variables

A linear combination of independent normally distributed random variables is itself normally distributed.

If $X_1 \sim N(\mu_1, \sigma_1^2)$, $X_2 \sim N(\mu_2, \sigma_2^2)$, ..., $X_n \sim N(\mu_n, \sigma_n^2)$ are independent and $Y = a_1X_1 + a_2X_2 + \cdots + a_nX_n$, where $a_1, a_2, \ldots, a_n$ are constants, then

$$Y \sim N(\mu, \sigma^2)$$

where
$$\mu = a_1\mu_1 + a_2\mu_2 + \cdots + a_n\mu_n$$
$$\sigma^2 = a_1^2\sigma_1^2 + a_2^2\sigma_2^2 + \cdots + a_n^2\sigma_n^2.$$

A special case, which is often used, is when $Y = X_1 - X_2$. In this case
$$\mu = \mu_1 - \mu_2$$
$$\sigma^2 = \sigma_1^2 + \sigma_2^2.$$

Exercise 10.4

1 The continuous random variables X and Y are independent, with X having the uniform distribution over the interval from 2 to 6, and with Y having the uniform distribution over the interval from 3 to 9. Assuming the results given in the information booklet for the mean and the variance of a uniform distribution, evaluate $E(XY)$ and $Var(3X - Y)$.   *(WJEC)*

2 The continuous random variable X is uniformly distributed between 9 and 21. The random variable Y has a Poisson distribution with mean 4. Given that X and Y are independent, find the mean and standard deviation of
  (i) $X - Y$,
  (ii) $4X + Y$.   *(JMB)*

3 X and Y are independent normally distributed random variables such that X has mean 32 and variance 25, and Y has mean 43 and variance 96. Find
  (i) $P(X > 43)$,
  (ii) $P(X - Y > 0)$,
  (iii) $P(2X - Y > 0)$.   *(JMB)*

4 The time, in hours, required to roast a chicken of weight w kg is a normally distributed random variable having mean $(0.4w + 2.2)$ and standard deviation $0.05w$.
  (i)   Find, to three decimal places, the probability that at least $3\frac{1}{4}$ hours will be required to roast a chicken weighing 2 kg.
  (ii)   Find the weight in kg, to three decimal places, of a chicken for which there is a probability of 0.95 that it will require less than 4 hours to roast it.
  (iii)   Find, to three decimal places, the probability that a chicken weighing 4 kg will require at least half an hour longer to roast than a chicken weighing 2 kg.

Three chickens, two weighing 2 kg and one weighing 4 kg, are to be roasted successively in random order. Find, to three decimal places, the probabilities that
  (iv)   the first chicken roasted will require at least $3\frac{1}{4}$ hours.
  (v)   the total time required to roast all three chickens will be less than $9\frac{1}{2}$ hours.   *(WJEC)*

5 An athlete A is to compete in the triple jump in the school sports. When the triple jump is a valid one, the lengths jumped by A in the three phases of the triple jump may be assumed to be independent and normally distributed with means 4.8 m, 2.1 m, 5.5 m and standard deviations 0.2 m, 0.1 m, 0.2 m, respectively. The total length of a triple jump is the sum of the lengths jumped in the three phases.
  (i)   Find the mean and the standard deviation of the total length of A's triple jump when it is valid.

In the school sports, each competitor in the triple jump makes six attempts. It is known that A has a probability of 0.8 of making a valid jump on each of his six attempts and the school record for the total length of a triple jump is 12.7 m.

(ii) Show that the probability that A will break the school record on his first attempt is 0.127, to three decimal places.

(iii) Find, to two decimal places, the probability that A will break the existing school record at least once during the school sports.

(iv) Comment briefly upon the assumption of independence made in this question. (*JMB*)

10.5 Sampling distributions

Sampling from a finite population

A simple random sample of size n from a finite population is a selection of n members of the population made in such a way that all possible selections of n members are equally likely to be chosen.

This definition covers sampling both with and without replacement.

Sampling from a distribution

A random sample of size n from the population distribution of X is a set of independent random variables $X_1, X_2, \ldots, X_n$, each of which has the same distribution as X.

This definition includes random sampling with replacement as a special case; it does not include random sampling without replacement.

Statistics

A statistic is a quantity calculated from the sample data for the purpose of estimating a population parameter.

When all possible random samples of a fixed size are taken from a population, the values of the statistic, which varies from one sample to another, form a distribution called the sampling distribution of the statistic.

Sample mean

If $\bar{X}$ is the sample mean of a random sample of size n taken from the distribution of X, which has mean μ and variance σ^2, then

$$E[\bar{X}] = \mu \quad \text{and} \quad V[\bar{X}] = \frac{\sigma^2}{n}.$$

This result is true whatever the distribution of X.

In addition, when the distribution of X is known to be normal, the distribution of $\bar{X}$ is also normal (for all n, large or small).

Central Limit Theorem

If $X_1, X_2, \ldots, X_n$ is a random sample of size n from any distribution having mean μ and variance σ^2 then, for large n, the distribution of the sample mean, $\bar{X} = \dfrac{(X_1 + X_2 + \cdots + X_n)}{n}$, is approximately normal with mean μ and variance $\dfrac{\sigma^2}{n}$.

For the approximation to be good, a sample size of 40 or more is usually large enough and when the population distribution is symmetrical, the approximation can be good for smaller sample sizes.

Under the same circumstances the distribution of the sum of the sample values, $T = X_1 + X_2 + \cdots + X_n$, is approximately normal with mean $n\mu$ and variance $n\sigma^2$.

Exercise 10.5

1 A bag contains 6 blue discs and 5 red discs. Three discs are randomly selected without replacement. Find the probability that the three selected discs

   (i) are all of the same colour,

   (ii) consist of two blue discs and one red disc.

   Find the mean and variance of the number of blue discs that will be selected.

   If instead the three discs are selected with replacement, write down the new mean and variance of the number of blue discs selected.

   Another bag contains 5 red discs and x blue discs. Two discs are randomly selected without replacement from this bag. Find the smallest value of x for which the probability that two blue discs will be selected is at least $\frac{1}{2}$.

   (JMB)

2 A cubical die has two each of its faces numbered 1, 2, and 3, respectively, and is such that the probabilities of obtaining these scores in a single throw are 0.1, 0.8, and 0.1, respectively.

   (i) If X is the score obtained in one throw of the die, determine the mean and the variance of X.

   (ii) Let M denote the median of the three scores obtained in three independent throws of the die. Show that $P(M = 1) = 0.028$. Evaluate $P(M = 2)$ and $P(M = 3)$, and hence determine the mean and the variance of the sampling distribution of M.

   (iii) Let $\bar{X}$ denote the mean of the three scores obtained in three independent throws of the die. Write down the values of the mean and the variance of $\bar{X}$, and verify that the variance of M is 84% of the variance of $\bar{X}$.

   (WJEC)

3 If $X_1, X_2, \ldots, X_n$ are independent Normally distributed random variables each having mean μ and variance σ^2, what is the distribution of

$$S_n = \sum_{i=1}^{n} X_i?$$

A doctor has found from experience that the times he takes to examine patients coming to his surgery are independent and Normally distributed with a mean of nine minutes and a standard deviation of three minutes. The doctor sees the patients consecutively with no time gaps between them. On a particular morning there are 16 patients in the surgery and the doctor sees the first patient at 9.00 a.m.

  (i)   Find the probability that exactly two of the first six patients will have examination times in excess of nine minutes.

 (ii)   Find the probability that the doctor will have examined all 16 patients by 11.30 a.m.

(iii)   Mrs. Jones, the ninth patient to be examined that morning, had pre-arranged to meet her husband outside the surgery. Find, to the nearest minute, the latest time at which the husband may turn up at the surgery if he is to have a probability of at least 0.99 of being there before his wife comes out.   (*WJEC*)

4 In a certain geographical region, the heights of girls and boys in the age range 16 to 18 years can be regarded as having normal distributions as follows.

Girls: mean 162.5 cm, standard deviation 4.0 cm.

Boys: mean 173.5 cm, standard deviation 4.8 cm.

Find the probability that

(i) a girl,   (ii) a boy   is taller than 170.5 cm.

A school sixth form in the area contains 64 girls and 36 boys in this age range. Regarding the girls and boys as random samples from the given normal populations, determine how many students in the sixth form would be expected to be taller than 170.5 cm.

Mean values of samples of size 64 and 36, drawn from the given normal populations of heights of girls and boys respectively, are denoted by $\bar{x}_G$ and $\bar{x}_B$ respectively.

State or calculate the means and variances of the sampling distributions of means $\bar{x}_G$ and $\bar{x}_B$ and of the sampling distribution of differences $\bar{x}_B - \bar{x}_G$. Deduce the probability that in the sixth form referred to above the mean height of the boys will be at least 12 cm more than that of the girls.

State briefly, giving a reason, whether or not it would be acceptable to regard the populations of heights of girls and boys aged 16 to 18 in this area as combining to form a single population of heights with a normal distribution.   (*JMB*)

5 In a large college the heights of the female students may be assumed to be normally distributed with mean 168 cm and standard deviation 8 cm, while the heights of the male students may be assumed to be normally distributed with mean 182 cm and standard deviation 6 cm. Calculate, to three decimal places in each case, the probabilities that

(i) a randomly chosen female student will be taller than a randomly chosen male student;

(ii) the mean of the heights of 15 randomly chosen female students and 10 randomly chosen male students will exceed 175 cm.

Given that 40% of all the students in the college are female, calculate, to three decimal places, the probability that a student chosen at random from all the students in the college will be taller than 188 cm. (*JMB*)

10.6 Estimation

A statistic T is said to be an unbiased estimator of a population parameter θ when $E[T] = \theta$. The value t of T is called an unbiased estimate of θ.

The standard deviation of the sampling distribution of a statistic T is often referred to as its standard error, $SE[T]$.

If T_1 and T_2 are two unbiased estimators of θ and $SE[T_1] < SE[T_2]$, or equivalently $V[T_1] < V[T_2]$, then T_1 is a better estimator of θ than T_2, since it is more likely to give an estimate which is close to θ.

The sample mean $\bar{X}$ is an unbiased estimator of the population mean μ and $SE[\bar{X}] = \dfrac{\sigma}{\sqrt{n}}$.

The sample proportion $\hat{P}$ is an unbiased estimator of the population

proportion p (or binomial probability parameter p) and $SE[\hat{P}] = \sqrt{\dfrac{p(1-p)}{n}}$.

The estimated value of this standard error is given by $ESE[\hat{P}] = \sqrt{\dfrac{\hat{p}(1-\hat{p})}{n}}$.

$$S^2 = \frac{\sum(X_i - \bar{X})^2}{n-1}$$

is an unbiased estimator for the population variance σ^2. The unbiased estimate s^2 may be evaluated using the formula

$$s^2 = \frac{\sum x_i^2 - n\bar{x}^2}{n-1}.$$

183

Exercise 10.6

1 The discrete random variable X has the distribution
$$P(X = 1) = 0.4, \qquad P(X = 2) = 0.2, \qquad P(X = 3) = 0.4.$$
Calculate the variance of X.

Let X_1 and X_2 denote two independent observations of X. Derive the sampling distribution of $T = \frac{1}{2}(X_1 - X_2)^2$, and verify that T is an unbiased estimator of the variance of X. *(WJEC)*

2 A bag contains 7 green balls and 3 red balls.

a From the bag 4 balls are selected at random *without* replacement. Let X denote the proportion of red balls in the sample.

(i) Show that
$$P(X = \tfrac{1}{4}) = 3P(X = 0).$$

(ii) Find the probability distribution of X.

(iii) Show that X is an unbiased estimator of the proportion p of red balls originally in the bag.

(iv) Find, to three decimal places, the standard error of X.

b From the original bag of 10 balls, 4 balls are selected at random *with* replacement. Let Y denote the proportion of red balls in the sample. Write down the mean of Y and evaluate, to three decimal places, its standard error.

c State, giving a reason, which of X and Y you consider to be the better estimator of p. *(JMB)*

3 The continuous random variable X has probability density function f, where
$$\text{f}(x) = 1, \qquad \theta + 1 \leqslant x \leqslant \theta + 2,$$
$$\text{f}(x) = 0, \qquad \text{otherwise},$$

where θ is an unknown constant. Denoting the mean and the variance of X by μ and σ^2, respectively, express μ in terms of θ, and find the value of σ^2.

A random sample of 10 observations of X had the values:
$$2.4, \quad 1.6, \quad 1.7, \quad 2.3, \quad 1.9, \quad 1.8, \quad 1.6, \quad 2.1, \quad 1.6, \quad 2.0.$$

Calculate an unbiased estimate of μ, and deduce an unbiased estimate of θ. Calculate, to three significant figures, the standard error of your unbiased estimate of θ.

Given that a random sample of 100 observations of X had mean 1.86, use a normal distribution approximation to calculate 95% confidence limits for θ, giving each limit correct to three significant figures. *(WJEC)*

4 A random sample of n observations is made of a random variable X which is uniformly distributed on the interval $(0, \theta)$. Write down an expression for the probability that all n observed values will be less than y. Hence deduce that the probability density of Y, the largest observation in a random sample of n values of X, is given by

$$g(y) = \frac{ny^{n-1}}{\theta^n}, 0 < y < \theta.$$

Show that $\dfrac{(n+1)Y}{n}$ is an unbiased estimator of θ and find the variance of this estimator.

5 A cubical die has two faces numbered 1, two numbered 2 and two numbered 3, and is such that when it is rolled the probability of scoring 1 is $\frac{1}{4}$, of scoring 2 is $\frac{1}{2}$ and of scoring 3 is $\frac{1}{4}$. If the die is rolled three times show that the probability that the smallest score will be 1 and the largest will be 3 is equal to $\dfrac{9}{32}$.

The average of the smallest and the largest values in a sample of n values is called the *midrange*. Find the sampling distribution of the midrange of the scores obtained when the above die is rolled three times. In this case show that the midrange is an unbiased estimator of the mean score per throw, and find the ratio of its sampling variance to that of the mean of the three scores. *(WJEC)*

10.7 Confidence intervals

A 95% confidence interval for a population parameter θ is an interval calculated by a method which is such that the probability of obtaining an interval which actually contains θ is 0.95.

1 Confidence limits for the mean of a normal distribution with known variance

$$\bar{x} \pm z_p \text{SE}[\bar{X}], \qquad \text{where } \text{SE}[\bar{X}] = \frac{\sigma}{\sqrt{n}}.$$

2 Approximate confidence limits for the mean of any distribution (large samples only)

$$\bar{x} \pm z_p \text{ESE}[\bar{X}], \qquad \text{where } \text{ESE}[\bar{X}] = \frac{s}{\sqrt{n}}.$$

3 Confidence limits for the difference between the means of two normal distributions with known variances

$$(\bar{x}_1 - \bar{x}_2) \pm z_p \text{SE}[\bar{X}_1 - \bar{X}_2],$$

$$\text{where } \text{SE}[\bar{X}_1 - \bar{X}_2] = \sqrt{\left(\frac{\sigma_1^2}{n_1} + \frac{\sigma_2^2}{n_2}\right)}.$$

4 Approximate confidence limits for the difference between the means of two distributions (large samples only)

$$(\bar{x}_1 - \bar{x}_2) \pm z_p\text{ESE}[\bar{X}_1 - \bar{X}_2],$$

$$\text{where ESE}[\bar{X}_1 - \bar{X}_2] = \sqrt{\left(\frac{s_1^2}{n_1} + \frac{s_2^2}{n_2}\right)}.$$

5 Approximate confidence limits for the population proportion or binomial probability parameter (large samples only)

$$\hat{p} \pm z_p\text{ESE}[\hat{P}], \qquad \text{where ESE}[\hat{P}] = \sqrt{\left[\frac{\hat{p}(1-\hat{p})}{n}\right]}.$$

6 Approximate confidence limits for the difference between two population proportions (large samples only)

$$(\hat{p}_1 - \hat{p}_2) \pm z_p\text{ESE}[\hat{P}_1 - \hat{P}_2],$$

$$\text{where ESE}[\hat{P}_1 - \hat{P}_2] = \sqrt{\left[\frac{\hat{p}_1(1-\hat{p}_1)}{n_1} + \frac{\hat{p}_2(1-\hat{p}_2)}{n_2}\right]}.$$

7 Approximate confidence limits for the mean of a Poisson distribution (large samples only)

$$\bar{x} \pm z_p\text{ESE}[\bar{X}], \qquad \text{where ESE}[\bar{X}] = \sqrt{\left(\frac{\bar{x}}{n}\right)}.$$

The values of z_p may be found from **Table 3**, (page 231).

8 Confidence limits for the mean of a normal distribution with unknown variance

$$\bar{x} \pm t_p\text{ESE}[\bar{X}], \qquad \text{where ESE}[\bar{X}] = \frac{s}{\sqrt{n}}.$$

The values of t_p may be found from **Table 4**, (page 232).

Exercise 10.7

1 The following table shows the observed frequency distribution of the number of matches per box in a random sample of 100 boxes of matches of a particular brand.

Number of matches	47	48	49	50	51	52	53
Number of boxes	3	8	16	30	24	15	4

Calculate unbiased estimates of the mean μ_1 and the variance of the number of matches per box of this brand. Determine an approximate 99% confidence interval for μ_1.

The numbers of matches per box in a random sample of 50 boxes of another brand gave unbiased estimates 49.6 and 2.17, respectively, of the mean μ_2 and the variance of the number of matches per box of this brand. Determine an approximate 90% confidence interval for $\mu_1 - \mu_2$.

2 The height x cm of each man in a random sample of 200 men living in the U.K. was measured. The following results were obtained:

$$\sum x = 35050, \qquad \sum x^2 = 6163\,109.$$

Calculate unbiased estimates of the mean and the variance of the heights of men living in the U.K.

Determine an approximate 90 per cent symmetric confidence interval for the mean height of men living in the U.K. Name the theorem which you have assumed.

A random sample of the heights of 400 women living in the U.K. gave unbiased estimates of 163.7 cm and 90.02 cm^2 for the mean and the variance, respectively, of the heights of women living in the U.K. Stating suitable null and alternative hypotheses, determine whether there is sufficient evidence at the 5 per cent significance level to support the claim that, for people living in the U.K., the mean height of men exceeds that of women by more than 10 cm. (*JMB*)

3 In an opinion poll taken in a large constituency, 1600 voters indicated their voting preference as follows:

Conservative 840;   Labour 760.

(i) Estimate the proportion of voters that will vote Conservative.

(ii) Based on this estimate, calculate the 95% symmetric confidence interval for the proportion of people that will vote Conservative.

(iii) Estimate the size of sample required to ensure that the 95% symmetric confidence interval for the proportion of people that will vote Conservative has range 0.02. (*JMB*)

4 Define the Poisson distribution. State its mean and variance. State under what circumstances the normal distribution can be used as an approximation to the Poisson distribution.

Readings, on a counter, of the number of particles emitted from a radioactive source in a time T seconds have a Poisson distribution with mean $250\,T$. A ten-second count is made. Find the probabilities of readings of

(i) more than 2600, (ii) 2400 or more.

A reading of 2000 is obtained, but the time T is not known. By considering a symmetrical two-sided 95% confidence interval, derive a quadratic equation whose roots are 95% confidence limits for T. (*JMB*)

5 When an object is weighed repeatedly using a certain type of balance, the observed masses are normally distributed with mean equal to the true mass of the object.

(i) Using a particular balance of the above type, it is known that the standard deviation of the observed masses is 0.5 mg.

    (*a*) If an object of true mass 5.0 mg is weighed four times on the balance calculate, correct to three decimal places, the probability

that the mean of the four observed weights will be greater than
5.1 mg.

(b) Find the least number of times that an object should be weighed
on this balance for the width of the 90% confidence interval for
the object's true mass to be less than 0.5 mg.

(c) An object was weighed 10 times on the balance and the mean of
the observed masses was 6.8 mg. Another object was weighed 15
times on the balance and the mean of the observed masses was
4.6 mg. Calculate 95% confidence limits for the difference
between the true masses of the two objects.

(ii) Using another balance of the above type, the standard deviation of
the observed masses is not known. Nine weighings of an object on
this balance gave the observed values (in mg):

$$7.4, \quad 7.5, \quad 7.5, \quad 7.4, \quad 7.4, \quad 7.6, \quad 7.5, \quad 7.5, \quad 7.7,$$

respectively. Calculate 95% confidence limits for the true mass of the
object.   (WJEC)

10.8 Hypothesis tests

The null hypothesis H_0 is an assumption made about a population parameter
θ in order to carry out the test. It takes the form $\theta = \theta_0$.

The alternative hypothesis H_1 is accepted when the result of the test indicates
that H_0 is unlikely to be true. The alternative hypothesis may be two-sided or
one-sided. When there is a prior indication or suspicion that θ may actually be
greater than the assumed value θ_0, then a one-tailed test is carried out with
H_1 in the form $\theta > \theta_0$; similarly, when there is a suspicion that θ may be less
than the assumed value, then a one-tailed test is carried out with H_1 in the form
$\theta < \theta_0$. Otherwise, a two-tailed test is carried out with H_1 in the form $\theta \neq \theta_0$.

The set of values of the test statistic which lead to the rejection of H_0 is called
the critical region.

The significance level of a test is the probability of rejecting the null hypothesis
when it is true.

The critical region of a two-tailed 5% significance test which uses the standard
normal variable as test statistic and $H_1: \theta \neq \theta_0$ is given by

$$(-\infty, -1.96) \cup (+1.96, +\infty).$$

The critical region of a one-tailed 5% significance test which uses the standard
normal variable as test statistic and $H_1: \theta > \theta_0$ is given by

$$(+1.645, +\infty).$$

The critical region of a one-tailed 5% significance test which uses the standard
normal variable as test statistic and $H_1: \theta < \theta_0$ is given by

$$(-\infty, -1.645).$$

The critical values for tests using the standard normal variable as test statistic
and having other significance levels may be found from **Table 3b** or **Table 3**
(page 231).

1 Test for the mean of a normal distribution with known variance σ^2

$$H_0: \mu = \mu_0, \text{ test statistic } Z = \frac{\bar{X} - \mu_0}{\frac{\sigma}{\sqrt{n}}} \sim N(0, 1).$$

2 Test for the mean of any distribution (large samples only)

$$H_0: \mu = \mu_0, \text{ test statistic } Z = \frac{\bar{X} - \mu_0}{\frac{S}{\sqrt{n}}} \text{ approximately } N(0, 1).$$

3 Test for the difference between two means of two independent normal distributions with known variances $\sigma_1{}^2, \sigma_2{}^2$

$$H_0: \mu_1 - \mu_2 = \theta_0, \text{ test statistic } Z = \frac{(\bar{X}_1 - \bar{X}_2) - \theta_0}{\sqrt{\frac{\sigma_1{}^2}{n_1} + \frac{\sigma_2{}^2}{n_2}}} \sim N(0, 1).$$

4 Test for the difference between two means of two independent distributions (large samples only)

$$H_0: \mu_1 - \mu_2 = \theta_0,$$

$$\text{test statistic } Z = \frac{(\bar{X}_1 - \bar{X}_2) - \theta_0}{\sqrt{\frac{S_1{}^2}{n_1} + \frac{S_2{}^2}{n_2}}} \text{ approximately } N(0, 1).$$

5 Test for a population proportion or binomial probability parameter

$$H_0: p = p_0$$

either (for small samples) test statistic $X \sim B(n, p_0)$
or (for large samples) test statistic X approximately $N(np_0, np_0(1 - p_0))$

$$\text{or test statistic } Z = \frac{\hat{P} - p_0}{\sqrt{\frac{p_0(1 - p_0)}{n}}} \text{ approximately } N(0, 1).$$

6 Test for the difference between two independent population proportions (large samples only)

$$H_0: p_1 - p_2 = \theta_0,$$

$$\text{test statistic } Z = \frac{(\hat{P}_1 - \hat{P}_2) - \theta_0}{\sqrt{\frac{\hat{P}_1(1 - \hat{P}_1)}{n_1} + \frac{\hat{P}_2(1 - \hat{P}_2)}{n_2}}} \text{ approximately } N(0, 1).$$

7 Test for the mean of a Poisson distribution

$$H_0: \alpha = \alpha_0,$$

either (for small α_0) test statistic $X \sim Po(\alpha_0)$
or (for large α_0) test statistic X approximately $N(\alpha_0, \alpha_0)$.

8   Test for the mean of a normal distribution with unknown variance

$$H_0: \mu = \mu_0, \text{ test statistic } T = \frac{\bar{X} - \mu_0}{\dfrac{S}{\sqrt{n}}} \sim t_{n-1}.$$

The critical values for this test are given in **Table 4**.

Type-1 and type-2 errors

P(type-1 error) is the probability of rejecting H_0 when H_0 is true. (α)
P(type-2 error) is the probability of not rejecting H_0 when H_0 is false. (β)
The power of a test is the probability of rejecting H_0 when it is false.
      Power $= 1 - \beta$

Exercise 10.8

1   The number of emissions from a radioactive source in a period of t minutes has a Poisson distribution with mean μt, where μ is unknown. Use the tables provided and a 5% significance level to test the null hypothesis $H_0: \mu = 2$ against the alternative hypothesis $H_1: \mu > 2$ in each of the cases

   (i)   when 15 emissions are observed in a period of 5 minutes;

   (ii) when 118 emissions are observed in a period of 50 minutes.   (*JMB*)

2   In n independent trials with probability p of success in each trial, show that the probability of r successes is

$$\binom{n}{r} p^r (1-p)^{n-r} \qquad (0 \leqslant r \leqslant n).$$

Show that the mean number of successes is np and state the value of the variance.

Mr. Smith claims that the probability, p, that any egg incubated by his hens will hatch is 0.7. A friend believes that p is less than 0.7, and ten eggs are used to test Mr. Smith's claim. Only three of the ten eggs hatch. Test whether this provides significant evidence that $p < 0.7$.

If 100 eggs are incubated, use the normal approximation to find the largest number r such that, if r or fewer hatch, the rejection of Mr. Smith's claim in favour of his friend's claim is justifiable at the five per cent significance level.   (*JMB*)

3   A machine is used to fill bags with potatoes. The weights of the potatoes in filled bags are known to be normally distributed with standard deviation 0.5 kg and with mean which depends on the machine setting.

   a   A random sample of four bags is taken when the machine is set to deliver 25 kg per bag. Given that the sample mean weight per bag is 24.6 kg, calculate a 95% confidence interval for the mean weight per bag. State whether or not your result is consistent with the mean weight per bag being 25 kg.

b It is suspected that the mean weight of potatoes per bag is somewhat less than the machine setting. To test this suspicion the machine is set to deliver 25 kg per bag. A random sample of 36 bags is then to be taken and the suspicion will be upheld only if the sample mean weight per bag is less than 24.9 kg.

    (i) Formulate this procedure in hypothesis testing terminology.

    (ii) Calculate the probability that the procedure will lead to a type-1 error.

    (iii) Calculate the power of the procedure if the mean weight delivered per bag is 24.85 kg. *(JMB)*

4 A factory produces large numbers of sweets in a variety of colours. Automatic machines select the sweets at random and pack them in boxes of 20. A random sample of 100 boxes was chosen, the contents of each box examined and the number of black sweets in each box recorded. The results obtained are summarised in the following table.

No. of black sweets	0	1	2	3	4	5	6 or more
No. of boxes	11	29	27	22	7	4	0

    (i) Find an unbiased estimate for the proportion p of sweets produced which are black, and, to three significant figures, an estimate of its standard error.

    (ii) Using a distributional approximation and a 5 per cent significance level, test the null hypothesis $p = 0.1$ against the alternative hypothesis $p \neq 0.1$. State your conclusion.

    (iii) Given that $p = 0.1$, use tables to find, to the nearest integer, the expected frequencies corresponding to the observed frequencies tabulated above. *(JMB)*

5 In an investigation into the effectiveness of a particular course in speed reading a group of 500 students was split into two groups, A and B, of sizes 300 and 200 respectively, thought to have been chosen at random.

Those in group A were given no special instruction; those in group B were given a course in speed reading. Each student was asked to read the same passage and the time taken was measured. The results were

Group A: mean time 78.4 s, variance $14 s^2$,

Group B: mean time 77.4 s, variance $15 s^2$.

Carry out a significance test to see if there is evidence that the course has improved reading speed. State carefully your null hypothesis, alternative hypothesis and final conclusion.

You learn later that, of the original 500 students, 200 students had decided for themselves that they wanted to take the course in speed reading and that these students became group B. Discuss briefly how this might affect your previous conclusion. *(JMB)*

10.9 Linear relationships

The least squares estimate for the true linear relation, $y = \alpha + \beta x$, between two algebraic variables x and y is given by

$$y = a + bx$$

where $\qquad b = \dfrac{\sum x_i y_i - n\bar{x}\bar{y}}{\sum x_i^2 - n\bar{x}^2} \qquad$ and $\qquad a = \bar{y} - b\bar{x}.$

When the x-values are accurate and the errors in the observed values of y are normally distributed with zero mean and variance σ^2.

1 Confidence limits for β

$$b \pm z_p \text{SE}[B], \qquad \text{where SE}[B] = \sqrt{\dfrac{\sigma^2}{\sum(x_i - \bar{x})^2}}$$

2 Confidence limits for α

$$a \pm z_p \text{SE}[A], \qquad \text{where SE}[A] = \sqrt{\dfrac{\sigma^2 \sum x_i^2}{n\sum(x_i - \bar{x})^2}}$$

3 Confidence limits for $y_0 = \alpha + \beta x_0$, the true value of y when $x = x_0$

$$\hat{y}_0 \pm z_p \text{SE}[\hat{Y}_0], \qquad \text{where SE}[\hat{Y}_0] = \sqrt{\dfrac{\sigma^2 \left[\sum(x_i - \bar{x})^2 + n(x_0 - \bar{x})^2 \right]}{n\sum(x_i - \bar{x})^2}}$$

In the above, $\hat{y}_0 = a + bx_0$ and $\sum(x_i - \bar{x})^2$ is evaluated using the formula

$$\sum(x_i - \bar{x})^2 = \sum x_i^2 - n\bar{x}^2$$

Regression

When an algebraic variable x and a random variable Y are such that

$$E[Y] = \alpha + \beta x$$

then the line $y = a + bx$ is called the least squares regression line of y on x. The same formulae are applicable, except that the confidence limits in **3** become confidence limits for the mean value of Y when $x = x_0$.

Exercise 10.9

1 In an investigation of the relationship $y = \alpha + \beta x$ connecting the two variables x and y, five experiments were conducted with x having the values 20, 30, 40, 50, and 60, respectively, and the corresponding values of y were measured. The least-squares estimate of the relationship was calculated from the results to be

$$y = 3.4 - 0.65x.$$

Assuming that the errors in the y-measurements are independent and normally distributed with mean zero and standard deviation 0.2, calculate 90% confidence limits for the values of

(i) α, \qquad (ii) β. \hfill (WJEC)

2   The relation between two quantities x and y is given by

$$y = \alpha + \beta x,$$

where α, β are unknown constants. Twenty pairs of values of x and y were obtained experimentally and the following results calculated:

$$\sum x = 40, \qquad \sum y = 263.4, \qquad \sum x^2 = 120, \qquad \sum xy = 725.4.$$

Assuming that the observed values of x are accurate and that the observed values of y are subject to independent random errors, each normally distributed with mean zero and standard deviation 0.2, determine

(i)   the least squares estimate for the equation $y = \alpha + \beta x$,

(ii)   the 99 per cent symmetric confidence interval for β.   (*JMB*)

3   The figures in the table give the wine consumption in the United Kingdom in millions of gallons (y) for the years 1963 to 1972 (x).

Year (x)	1963 1964 1965 1966 1967 1968 1969 1970 1971 1972
Consumption (y) (millions of gallons)	32.5 37.1 35.5 37.7 41.5 46.4 44.8 45.8 53.9 62.0

Draw a scatter diagram to show these data.

Determine the least squares estimate of the regression line of y on x, showing all your working. Draw this line on your scatter diagram and use it to estimate the consumption for 1973.

Comment on the appropriateness of a *linear* regression model in this case, given also that the actual wine consumption in 1973 was 78.3 million gallons.   (*JMB*)

4   An experiment was conducted to determine the mass y grams of a given amount of chemical that dissolved in glycerine at $x°C$. The results of the experiment are given in the following table.

Temperature ($x°C$)	0	10	20	30	40	50
Mass (y grams)	51.3	51.4	51.9	52.0	52.6	52.8

Assuming that the true value of y is linearly related to the value of x, obtain the least squares estimate of this relationship.

Assuming further that the temperatures used in the experiment were controlled accurately but that the measured values of y were subject to independent random errors which are normally distributed with mean zero and standard deviation 0.2 g, calculate 95% confidence intervals for

(i)   the mass of chemical that will dissolve in glycerine at $0°C$,

(ii)   the *additional* mass of chemical that will dissolve in glycerine when the temperature is raised from $10°C$ to $20°C$.   (*WJEC*)

5   It is known that the true response, y, in a certain chemical process is linearly related to the operating temperature $x°$ C. However, experimental determinations of y are subject to random errors, so that when the process is run at temperature $x_i°$C the observed response is given by

$$y_i = \alpha + \beta x_i + e_i,$$

where $\alpha + \beta x_i$ is the true response and e_i is the error. The following table gives the responses that were observed in nine runs, three at each of the temperatures 20°C, 30°C, and 40°C.

Temperature (x_i)	20	30	40
Observed responses (y_i)	35	31	28
	33	32	27
	34	31	29

(i)   Calculate the least squares estimates of α and β.
(You may assume that the sum of the products, $\sum x_i y_i$, of the 9 pairs of observed values (x_i, y_i) is equal to 8220.)

(ii)   The errors, e_i, are independent and normally distributed with mean zero and standard deviation 2.5. Calculate 95% confidence intervals for
(a)   the value of β,
(b)   the true value of y when $x = 40$.

(iii)   Explain why the confidence interval you calculated in (ii)(b) is preferable to the one that could be obtained from the three observed responses when $x = 40$.   (WJEC)

Miscellaneous Exercise ☐ 10 ☐

1   The following table gives a grouped frequency distribution of the annual salaries of the 50 employees of a small company.

Salary (in £)	4000–	5000–	6000–	7000–	8000–	10000–	12000–	20000–
Frequency	5	11	12	10	6	4	2	0

(i)   Draw a histogram on graph paper to illustrate these data.
(ii)   Obtain, showing your methods, estimates of the mean and the median of the salaries.
(iii)   State one advantage, in this case, of either of these measures of central location.   (JMB)

2   (i)   In 1983 the mean and variance of the numbers of years of service of six men in a firm were 7 and 22, respectively; the mean and variance of the numbers of years of service of four women in the firm were 5 and 12.5, respectively. Find the mean and the variance of the numbers of years of service of these ten employees.

(ii)   Three years later in 1986, the firm made these ten employees redundant and paid each of them a redundancy payment of £100 for each year of service. Calculate the mean and, to the nearest £, the standard deviation of these redundancy payments.   (*JMB*)

3 The continuous random variables X and Y are independent. X is uniformly distributed between 7 and 19, and Y is normally distributed with mean 12 and variance 48.

(i)   Find the mean and the variance of $X - Y$, indicating where the independence of X and Y is relevant.

(ii)   Find the two values of the constant a for which the variance of aX is equal to the variance of Y.   (*JMB*)

4 The two events A and B are such that
$$P(A) = 0.2, \qquad P(B) = 0.3, \qquad \text{and} \qquad P(A \cup B) = 0.4.$$

(i)   Determine whether or not A and B are independent.

(ii)   Evaluate $P(B|A')$.   (*WJEC*)

5 Two events A and B are such that $P(A) = 0.2$ and $P(A \cup B) = 0.7$. Find $P(B)$, in each of the following cases,

(i)   when A and B are mutually exclusive,

(ii)   when A and B are independent.   (*JMB*)

6 Three events A, B, C are such that B and C are independent, C and A are independent, A and B are independent and
$$P(A) = \frac{1}{2}, \qquad P(B) = \frac{1}{3}, \qquad P(C) = \frac{1}{4}, \qquad P(A' \cap B' \cap C') = \frac{5}{24}.$$

(i)   Find $P(A \cup B)$.

(ii)   Find $P(A \cup B \cup C)$ and show that $P(A \cap B \cap C) = \frac{1}{12}$.   (*JMB*)

7 Carriers of a rare but deadly disease form 0.05 *per cent* of the whole population. A test has been developed to detect carriers of this disease. The test always gives a positive response when it is applied to carriers of the disease, but it also gives a positive response with probability 0.01 when it is applied to non-carriers.

(i)   Calculate the probability that the test will give a positive response when it is applied to a person selected at random from the whole population.

(ii)   The test is applied to a person selected at random. Given that the response is positive, calculate, to four decimal places, the conditional probability that the person tested is a carrier of the disease.

(iii)   State briefly a possible disadvantage of this test for mass screening purposes.   (*JMB*)

8 Two archers A and B shoot alternately at a target until one of them hits the centre of the target and is declared the winner. Independently, A and B have probabilities of $\dfrac{1}{3}$ and $\dfrac{1}{4}$, respectively, of hitting the centre of the target on each occasion they shoot.

   a Given that A shoots first, find

      (i)   the probability that A wins on his second shot,

      (ii) the probability that A wins on his third shot,

      (iii) the probability that A wins.

   b Given that the archers toss a fair coin to determine who shoots first, find the probability that A wins.   (*JMB*)

9 In each trial of a random experiment, the probability that the event A will occur is 0.6. In eight independent trials of the experiment calculate, correct to four decimal places, the probabilities that

   (i) A will occur in exactly five of the trials,

   (ii) A will occur in exactly five of the trials, these trials being consecutive.   (*WJEC*)

10 Four cards are numbered 2, 2, 4 and 6, respectively. Two of these cards are chosen at random without replacement. Let Y denote the larger of the numbers on the two chosen cards. (Take $Y = 2$ if the two chosen cards are both numbered 2.) Determine the sampling distribution of Y and hence find the expected value of Y.   (*WJEC*)

11 A discrete random variable X can take only the values 0, 1, 2 or 3, and its probability distribution is given by

$$P(X = 0) = k, \quad P(X = 1) = 3k, \quad P(X = 2) = 4k, \quad P(X = 3) = 5k,$$

where k is a constant. Find

   (i)   the value of k,

   (ii) the mean and variance of X.   (*JMB*)

12 A planted tuber produces 0, 1, or 2 shoots with probabilities 0.2, 0.3 and 0.5, respectively. Independently for each shoot produced, the probability that it will flower is 0.4. Let X denote the number of shoots from a planted tuber that will flower.

   (i)   Show that $P(X = 0) = 0.56$ and that $P(X = 1) = 0.36$.

   (ii) Find the mean and the standard deviation of X.   (*WJEC*)

13 The continuous random variable X has probability density function f given by

$$\begin{aligned} f(x) &= 3x^2, &&\text{for } 0 < x < 1, \\ f(x) &= 0, &&\text{otherwise.} \end{aligned}$$

Find the mean and the variance of

   (i) X,   (ii) X^{-1}.   (*WJEC*)

14 The length of each side of a cube is X cm, where X is a continuous random variable having probability density function f given by

$$f(x) = \frac{2}{x^2}, \qquad \text{for } 1 \leqslant x \leqslant 2,$$

$$f(x) = 0, \qquad \text{otherwise.}$$

Find the mean and the variance of the volumes of the cubes. *(WJEC)*

15 The distribution function F of a continuous random variable X is given by

$$F(x) = 0, \qquad\qquad x < 0,$$

$$F(x) = \frac{x^2}{48}(16 - x^2), \qquad 0 \leqslant x \leqslant 2,$$

$$F(x) = 1, \qquad\qquad x > 2.$$

Given that $Y = 16 - X^2$, show that the probability density function g of Y is given by

$$g(y) = \frac{1}{24}(y - 8), \qquad 12 \leqslant y \leqslant 16,$$

$$g(y) = 0, \qquad\qquad \text{otherwise.}$$

Find $E(Y)$. Hence, or otherwise, find $E(X^2)$. *(JMB)*

16 The probability density function f of a continuous random variable X is given by

$$f(x) = \ln x, \qquad 1 \leqslant x \leqslant e,$$

$$f(x) = 0, \qquad \text{otherwise.}$$

(i) Show that the mean, μ, of X is $\frac{1}{4}(e^2 + 1)$.

(ii) Find $P(X < \mu)$, giving your answer correct to two decimal places.

(iii) Show that μ is less than the median value of X. *(JMB)*

17 The continuous random variable X, which is restricted to values in the interval from 2 to 4, inclusive, has cumulative distribution function F given by

$$F(x) = ax^2 + bx, \qquad \text{for } 2 \leqslant x \leqslant 4.$$

(i) Giving a clear indication of your method, show that $a = \frac{1}{8}$ and $b = -\frac{1}{4}$.

(ii) Find the value of c such that $P(X > c) = 0.88$.

(iii) Find the cumulative distribution function of $Y = X(X - 2)$.

(iv) Let N denote the integer obtained by rounding an observed value of X to its nearest integer value. Find the probability distribution of N and, hence, evaluate $E[N]$. *(WJEC)*

18 The continuous random variable X has probability density function f, where

$$f(x) = \frac{25}{12(x+1)^3}, \qquad \text{for } 0 \leqslant x \leqslant 4,$$

$$f(x) = 0, \qquad \text{otherwise.}$$

(i) Evaluate $E[X + 1]$. Hence, or otherwise, find the mean of X.

(ii) Find the value of $c > 0$ for which $P(X \leqslant c) = c$.

(iii) Find the cumulative distribution function of $Y = (X + 1)^{-2}$.

Hence, or otherwise, find the probability density function of Y.   (*WJEC*)

19 Independently on each day, the number of times that a machine will require attention has the Poisson distribution with mean 1.2. Find, correct to three decimal places, the probabilities that

(i) on any day, the machine will require attention exactly three times.

(ii) on two successive days, the machine will require attention twice on one of the days and once on the other day.

The cost of running the machine on a day when it requires attention X times is £$(12 + 10X^2)$.

(iii) Find the expected daily cost of running the machine.   (*WJEC*)

20 Two types of flaw, A and B, may occur in manufactured cloth. The number of type A flaws per hundred metres of cloth has a Poisson distribution with mean 1 and, independently, the number of type B flaws per hundred metres of cloth has a Poisson distribution with mean 0.5. The cost of rectifying a type A flaw is 7 pence and that of rectifying a type B flaw is 12 pence.

(i) Calculate, to three decimal places, the probability that the cost of rectifying the flaws in one hundred metres of cloth will be less than 13 pence.

(ii) Calculate the mean and the standard deviation of the cost of rectifying flaws per one hundred metres of cloth.   (*JMB*)

21 The daily demand X for a particular video in a certain shop has a Poisson distribution with mean 3.5.

(i) Find the least number of copies of the video that the shop should stock in order that the probability of meeting the demand should exceed 0.95.

(ii) If the shop has five copies of the video available for daily hire, find the probability distribution of Y, the number of copies hired out daily by the shop, giving the probabilities to three decimal places.

Calculate the expected value of Y, giving your answer to two decimal places.   (*JMB*)

22 In a certain factory the number of accidents per shift was recorded for a period of 100 consecutive shifts and the following results were obtained.

No. of accidents	0	1	2	3	4	5	6	7 or more
No. of shifts	14	28	25	18	9	4	2	0

Show that the mean of this distribution is 2 and find (showing your method) the variance.

State why it would be reasonable to use the Poisson distribution with mean 2 as a model for the distribution of the number of accidents per shift in this factory. Using this model, find, correct to the nearest integer, the expected frequencies of the number of accidents per shift in a period of 100 shifts.

New working practices are to be introduced and the following test is to be applied to investigate whether the accident rate per shift has been increased as a result.

If there are three or more accidents on each of the 25th, 50th and 75th shifts after the introduction of the new practices, then this will be accepted as evidence that the accident rate has been increased.

Assuming that the mean number of accidents per shift before the introduction of the new practices was 2, write down appropriate null and alternative hypotheses and calculate the probability of a type-1 error, giving your answer to three decimal places. (*JMB*)

23 The probability that a clover has four leaves is 0.002.
  (i) Calculate, to three significant figures, the probability that a random sample of 5 clovers will include exactly one having four leaves.
  (ii) Find the smallest number of clovers that should be sampled if there is to be a probability of at least 0.9 that the sample will include one or more having four leaves.
  (iii) Use one of the tables provided to find an approximate value for the probability that a random sample of 1000 clovers will include at least 2 having four leaves.
  (iv) Each of 50 persons takes a random sample of 1000 clovers. Find an approximate value for the probability that exactly 25 of the persons will each have at least two clovers having four leaves. (*WJEC*)

24 The manufacturer of a fly spray claims that it kills 95 per cent of all flies sprayed.
  a Assuming that the manufacturer's claim is true, find, to three decimal places,
    (i) the probability that exactly 10 flies are killed when 12 flies are sprayed,
    (ii) the probability that at least 17 flies are killed when 20 flies are sprayed,
    (iii) an approximate value for the probability that at least 95 flies are killed when 100 flies are sprayed.

b When a random sample of 500 flies are sprayed a total of 470 flies were killed. Stating suitable null and alternative hypotheses, determine whether there is sufficient evidence at the 5 per cent significance level to conclude that the spray kills less than 95 per cent of flies sprayed. Explain the meaning of the term '5 per cent significance level'. (*JMB*)

25 A self-service cafeteria offers only two main courses, *A* and *B*, at lunchtime. Experience has shown that 60% of the customers opt for *A*.

  (i) Calculate the probabilities that on any day
    (*a*) exactly 3 of the first 6 customers will opt for *A*,
    (*b*) the fourth customer will be the first to opt for *B*.

  (ii) Using the tables provided find, to four decimal places, the probability that of the first 10 customers on any day, 7 or fewer will opt for *A*.

  (iii) On a particular day only 30 servings of *A* and 25 servings of *B* are available. Use the tables provided to find, to four decimal places, the probabilities that of the first 50 customers on that day,
    (*c*) every one will be able to have the course of his/her choice,
    (*d*) the number opting for *A* will be at least twice the number opting for *B*.

  (iv) On a day when the cafeteria had an unlimited supply of each course, there were 120 customers. Use a distributional approximation to calculate, to three decimal places, the probability that at least 80 of the 120 customers opted for *A*. (*WJEC*)

26 A box contains ten coins. Four of the coins are fair, while each of the other six coins is such that when tossed the probability of obtaining a head is $\frac{1}{4}$.

  (i) One coin is chosen at random from the box and is tossed twice. Let *A* denote the event that the first toss gives head, and let *B* denote the event that the second toss gives head. Show that *A* and *B* are not independent.

  (ii) Suppose, instead, that two coins are chosen at random from the ten coins in the box and that they are tossed together once.
    (*a*) Calculate the probability that one head and one tail will be obtained.
    (*b*) Given that one head and one tail were obtained, calculate the conditional probability that at least one of the two chosen coins was a fair coin. (*WJEC*)

27 A box labelled *A* contains three white balls and five black balls. Three balls are drawn at random without replacement from the box and placed in an empty box labelled *B*.

  (i) Find the probability distribution of the number of white balls drawn from box *A*.

One ball is now drawn at random from the five balls remaining in box A, and one ball is drawn at random from the three in box B.

(ii) Calculate the probability that the ball drawn from B is white.

(iii) Given that the ball drawn from A and the ball drawn from B are the same colour, calculate the probability that the three balls transferred from A to B consisted of one white ball and two black balls.   (*WJEC*)

28 Suppose that the heights of adult males are normally distributed with mean 174.5 cm and standard deviation 10 cm.

(i) Calculate, correct to three decimal places, the probability that the height of a randomly chosen adult male will be greater than 175 cm.

(ii) Use a distributional approximation to calculate the probability that at least 40 of 80 randomly chosen adult males will have heights greater than 175 cm; give your answer correct to three decimal places.   (*WJEC*)

29 A soft-drinks machine is regulated so as to deliver an average of 200 ml per cup. The actual amount delivered per cup is a normally distributed random variable having mean 200 ml and standard deviation 8 ml.

(i) Given that the maximum capacity of a cup is 210 ml find, to three decimal places, the probability that a cup will overflow.

(ii) Find, to the nearest ml, what the maximum capacity of each cup should be for only 1% of the cups to overflow.   (*WJEC*)

30 A garage sells three grades of petrol, namely 2-star, 3-star, and 4-star. In a week the amounts in litres sold of 2-star, 3-star, and 4-star are independent and normally distributed with means 7000, 13 000, and 22 000, respectively, and standard deviations 320, 400, and 500, respectively.

(i) Calculate, to three decimal places in each case, the probabilities that in a week
   (a) the combined total amount of petrol sold will exceed 41 000 litres,
   (b) the amount of 4-star sold will exceed the combined amounts of 2-star and 3-star sold.

(ii) Find the value of c, to the nearest litre, if the combined total amount sold exceeds c litres in 90% of the weeks.

(iii) The profits per litre sold are 1.5 pence for 2-star petrol and 2.5 pence for 4-star petrol. Calculate, to three decimal places, the probability that in a week the profit from the sale of 4-star petrol is at least five times that from the sale of 2-star petrol.   (*WJEC*)

31 A bag contains 2 white balls and 3 black balls. Two balls are drawn simultaneously and at random from the bag. Each black ball drawn (if any) is then replaced in the bag, and each white ball (if any) is discarded. Two balls are then drawn simultaneously and at random from the balls that are now in the bag. Let X denote the number of white balls obtained in the first draw, and let Y denote the number of white balls obtained in the second draw.

(i)   Show that $P(X = 1, Y = 1) = 0.3$, and that $P(X = 2, Y = 0) = 0.1$.

(ii) Construct a two-way table for the joint probability distribution of X and Y.

(iii) Find the expected value of Y.

(iv) Each white ball obtained in either draw scores 3 points, while each black ball obtained in either draw scores 1 point. Let T denote the sum of the scores after the two draws. Find the mean and the variance of T. *(WJEC)*

32 The discrete random variables X and Y are independent with probability distributions as shown in the following tables.

Distribution of X

x	1	2
$P(X = x)$	0.4	0.6

Distribution of Y

y	1	2	3
$P(Y = y)$	0.3	0.4	0.3

(i)   Evaluate $E\left(\dfrac{Y}{X}\right)$.

Let $U = XY$ and $W = 2X - Y$.

(ii) Show that $E[UW] = E[U]E[W]$.

(iii) Evaluate $P(U = 2, W = 0)$.

(iv) Determine whether or not U and W are independent. *(WJEC)*

33 The lifetime, x thousand hours, was recorded for each electric light bulb in a random sample of 500 taken from the production of factory A. The following results were calculated:

$$\sum x = 525.50, \qquad \sum x^2 = 565.0749.$$

Calculate unbiased estimates for the mean μ and the variance of the lifetimes of bulbs produced by factory A. Use your results to calculate an approximate 95 per cent confidence interval for μ.

A random sample of 400 electric light bulbs produced by factory B gave unbiased estimates of 1.072 and 0.0196, respectively, for the mean and the variance of the lifetimes (in thousands of hours) of bulbs produced by factory B. Stating suitable null and alternative hypotheses, determine whether there is sufficient evidence at the 5 per cent significance level to support the hypothesis that the mean lifetimes of the bulbs produced by factories A and B differ. *(JMB)*

34 A cable hanging vertically from a fixed point supports a load; the relation between the length, y metres, of the cable and the load, x kilograms, is given by $y = \alpha + \beta x$, where α, β are unknown constants. In an experiment to estimate α and β, fifteen pairs of values of x and y were recorded and the following results calculated:

$$\sum x = 195, \quad \sum y = 226.80, \quad \sum x^2 = 2815, \quad \sum xy = 2951.76.$$

The observed values of x are accurate and the observed values of y are subject to independent random errors, each normally distributed with zero mean and standard deviation 0.01 metres.

(i) Determine the least squares estimate for the equation
$$y = \alpha + \beta x.$$

(ii) Calculate the 90 per cent symmetric confidence limits for β, giving your answers to three decimal places.

(iii) Deduce the 90 per cent symmetric confidence interval for the increase in the value of y when x is increased from 10 to 15 kilograms. (*JMB*)

35 An investigation of a possible linear relationship between two variables x and y was conducted with x having five prespecified values, the corresponding values of y being observed. The results obtained are given in the following table:

x	5	10	15	20	25
y	55	52	50	48	45

You may assume that $\sum x^2 = 1375$ and that $\sum xy = 3630$.

Given that the true relationship between x and y is $y = \alpha + \beta x$, determine the least-squares estimates of α and β.

Suppose that the observed values of y are subject to independent random errors that are normally distributed with a mean of zero and a standard deviation of 0.5. Calculate 90 per cent confidence limits for

(i) the true value of y when $x = 20$,

(ii) the difference between the true values of y when $x = 5$ and $x = 25$, respectively. (*WJEC*)

36 The size (z) of an organism was measured at different times (x) giving the data shown in the first two rows of the following table. The third row of the table gives the values of $y = \log_{10} z$ to two decimal places. Without carrying out any nontrivial calculations show that the assumption of a linear relationship between y and x is more realistic than between z and x.

Time (x)	1	3	5	7	9
Size (z)	1.48	3.02	5.37	9.55	17.38
$y = \log_{10} z$	0.17	0.48	0.73	0.98	1.24

The following quantities were calculated from the above data:

$$\sum x = 25, \qquad \sum x^2 = 165, \qquad \sum y = 3.6, \qquad \sum xy = 23.28$$

Suppose that $y = \alpha + \beta x$ and that for given values of x the observed values of y are subject to independent errors which are Normally distributed with mean zero and standard deviation 0.02.
(i) Calculate the least squares estimates of α and β.
(ii) Obtain 95 per cent confidence limits for the value of y when $x = 10$. Hence find 95 per cent confidence limits for the size of the organism at time $x = 10$. (*WJEC*)

37 It is known that the true response y in a certain chemical experiment is a linear function of the operating temperature x. However, the experimental determinations of y are subject to random errors, so that when an experiment is performed at temperature x_i the observed response y_i is such that
$$y_i = \alpha + \beta x_i + e_i,$$
where $\alpha + \beta x_i$ is the true response and e_i is the error. The following table gives the observed responses in six experiments, two at each of three temperatures.

Temperature (x)	30	40	50
Observed responses (y)	14	10	7
	12	11	6

Use the data to obtain the least squares estimate of the linear relationship connecting y and x. [You are given that $\sum xy = 2270$.]

The errors, e_i, are independent and Normally distributed with zero mean and unit standard deviation.

Calculate 90 per cent confidence limits for
(i) the value of α,
(ii) the value of β,
(iii) the true value of y, when $x = 50$. (*WJEC*)

38 The random variables X and Y are independent, with X having mean μ and variance σ^2, and Y having mean μ and variance $\dfrac{\sigma^2}{k}$, where k is a positive constant. Let $\bar{X}$ denote the mean of a random sample of 10 observations of X, and let $\bar{Y}$ denote the mean of a random sample of 15 observations of Y. Show that
$$T_1 = \frac{2\bar{X} + 3k\bar{Y}}{2 + 3k} \qquad \text{and} \qquad T_2 = \frac{2\bar{X} + 3\bar{Y}}{5}$$
are both unbiased estimators of μ.

Find expressions for the variances of T_1 and T_2, and show that the variance of T_1 is less than or equal to the variance of T_2.

Find, in terms of k, the value a for which
$$T = a\bar{X} + (1 - a)\bar{Y}$$
has smallest possible variance. (*WJEC*)

39 The number of air bubbles in the glass of a bottle has a Poisson distribution with mean μ. Bottles containing one or more air bubbles are set aside as being unsatisfactory. Let X denote the number of air bubbles in an *unsatisfactory* bottle. Show that

$$P(X = r) = \frac{\mu^r}{(e^\mu - 1)r!}, \text{ for } r = 1, 2, 3,\ldots$$

Find an expression for the mean number of air bubbles per *unsatisfactory* bottle.

In a random sample of n unsatisfactory bottles, let $\bar{X}$ denote the sample mean number of air bubbles per bottle, and let Y denote the number of the n bottles that contain exactly one air bubble.

Show that $\bar{X} - \dfrac{Y}{n}$ is an unbiased estimator of μ. (*JMB*)

40 Let $\bar{X}$ denote the mean of a random sample of 25 observations from a normal distribution whose mean and standard deviation are both equal to μ. Find, to two decimal places, values for a and b if the interval $(a\bar{X}, b\bar{X})$ is to have probability 0.95 of including the value of μ. (*JMB*)

41 A botanist measures the lengths of a random sample of leaves of a certain variety of tree. The botanist states that 'the symmetric 95% confidence limits for the mean length of such leaves are 5.68 cm and 7.14 cm'. Explain carefully what is meant by this statement.

Assuming that the lengths of the leaves are normally distributed with mean μ cm and standard deviation 2 cm, calculate the number of leaves in the botanist's sample and the mean length of the sampled leaves.

The same botanist also measures the lengths of a random sample of m leaves from a different variety of tree. The mean length of these sampled leaves is 5.47 cm. It may be assumed that the lengths of leaves of this variety are normally distributed with mean λ cm and standard deviation 2 cm. The botanist then proceeds to test, at the 1% significance level, the null hypothesis that $\mu = \lambda$.
(i) If the alternative hypothesis is $\mu \neq \lambda$, show that the null hypothesis will not be rejected whatever the value of m.
(ii) If the alternative hypothesis is $\mu > \lambda$, find the smallest value of m for which the null hypothesis will be rejected. (*JMB*)

42 Every schoolday a teacher drives 30 kilometres to school. The time, X hours, taken to complete the journey on any day is a continuous random variable having probability density function f, where

$$f(x) = 12x(1 - x), \qquad \tfrac{1}{2} \leqslant x \leqslant 1,$$
$$f(x) = 0, \qquad \text{otherwise.}$$

Show that the expected value of $\dfrac{1}{X}$ is equal to 1.5.

Find the mean and, correct to two decimal places, the standard deviation of the teacher's average speed for the journey to school. (*JMB*)

43 At a certain garage the times, in minutes, taken to complete a standard service on each of a random sample of 10 cars were

66, 88, 74, 79, 87, 98, 83, 75, 86, 68.

(i) Write down an unbiased estimate of the proportion of such services that are completed in under 85 minutes and calculate, correct to three decimal places, an estimate of its standard error.

(ii) Calculate unbiased estimates of the mean and the variance of all the times taken for a standard service at this garage.

Assuming that the service times are normally distributed with standard deviation 10 minutes, calculate 95% confidence limits for the mean service time, giving your answers correct to one decimal place. (*JMB*)

44 The probability that a seed of a particular variety will germinate is p, where p is unknown. In a random sample of 20 such seeds, let X be the number of seeds that will germinate. The null hypothesis that $p = 0.6$ is to be tested.

a Given that the null hypothesis is to be rejected if $X \leqslant 8$ or $X \geqslant 16$, find

(i) the significance level of the test,

(ii) the probability of a type-2 error when $p = 0.4$.

b Given that the null hypothesis is to be rejected if $X \leqslant 12 - r$ or $X \geqslant 12 + r$, find the value of r for which the test has a significance level of approximately 0.01. (*JMB*)

45 The length, X cm, of an offcut of wooden planking is a random variable which has a rectangular distribution defined on the interval $(0, \theta)$, where θ is unknown.

State, in terms of θ, the mean and variance of X.

Hence show that, if the random variable $\bar{X}$ denotes the mean length of a random sample of n offcuts, then $2\bar{X}$ is an unbiased and consistent estimator of θ.

Explain what is meant by the *sampling distribution of an estimator*.

State, with a justification, the approximate sampling distribution of $2\bar{X}$ when n is large.

An analysis of the lengths of a random sample of 100 offcuts gave $\sum x = 1050$ cm.

(i) Estimate the value of θ.
    Hence estimate the probability that the length of a randomly selected offcut exceeds 15 cm.

(ii) Calculate an approximate 95% symmetric confidence interval for θ.
(*JMB*)

46 The number of toasters sold per day by a shop has a Poisson distribution with mean 4.5. Independently, the number of kettles sold per day by the shop has a Poisson distribution with mean 3.

(i) Find, to two decimal places, the probability that the number of toasters that will be sold in a day will be between 4 and 6, inclusive.

(ii) Without using tables, show that, on a randomly chosen day, the probability that the shop will sell one toaster and one kettle is three times the probability that the shop will sell no toasters and two kettles.

The shop makes a profit of £2 on each toaster sold and a profit of £3 on each kettle sold.

(iii) Find the mean and the variance of the shop's total daily profit from the sale of toasters and kettles.

(iv) Using a normal approximation and ignoring the correction for continuity find, to two decimal places, the probability that over a period of 40 days the shop's mean daily profit from the sale of toasters and kettles will exceed £20.

(v) Given that, on a particular day, the shop's total profit from the sale of toasters, and kettles was £10, calculate, to two decimal places, the probability that the profit from the sale of toasters was greater than that from the sale of kettles. (*JMB*)

47 Explain what is meant by the *null hypothesis* and the *alternative hypothesis* in significance testing.

Prior to joining a training course, a netball player has scored with half of her shots at the basket and is thus considered to have a scoring ability of 0.50. After completing the training course, which involved coaching in shooting skills, and before her next game, she claims that her shooting ability has improved as a result of receiving this coaching.

State suitable null and alternative hypotheses to test her claim.

Her claim will be accepted if, in her next game, she scores with at least 10 of her 12 shots at the basket. Assuming her shots to be independent, find the significance level of the test.

Over the next few games she scores with 107 of her 125 shots at the basket. Using the normal approximation to the binomial distribution, test, at the 1% level of significance, her coach's claim that, after attending the training course, her scoring ability is greater than 0.75. (*JMB*)

48 Explain briefly what is meant by a *random sample*.

The alkalinity, in milligrams per litre, of the water in a river at a point U, which is 10 km upstream of a factory's discharge pipe, is known to be normally distributed with a mean of 75 mg/l and a standard deviation of 10 mg/l.

Determine the probability that

(i) a single alkalinity measurement at U exceeds 90 mg/l.

(ii) the mean of a random sample of 10 alkalinity measurements at U is less than 70 mg/l.

To assess the effect of the factory's discharge of waste water on the quality of the water in the river, 16 samples of water are taken from the river at a point D, which is 5 km downstream of the factory's discharge pipe. The alkalinity, x milligrams per litre, is measured for each sample with the following results.

136	86	113	95	108	79	100	121
86	91	97	106	113	121	140	136

$$(\Sigma x = 1728 \qquad \Sigma x^2 = 192060)$$

Assuming that these alkalinity measurements are also normally distributed but not necessarily with a standard deviation of 10 mg/l, test, at the 5% level of significance, the hypothesis that the population mean alkalinity of the water at D exceeds that at U by 25 mg/l.   (JMB)

49 In a survey the durations, in minutes, of telephone calls were measured for a random sample of 1000 local calls made at peak rate. The results, given as percentages, are summarized in the table below.

Duration, x mins	Percentage
$0 \leqslant x < 1$	14
$1 \leqslant x < 3$	22
$3 \leqslant x < 10$	42
$10 \leqslant x < 15$	10
$15 \leqslant x < 30$	9
$30 \leqslant x < 60$	3

The diagram below was used in the report of the survey to illustrate the pattern.

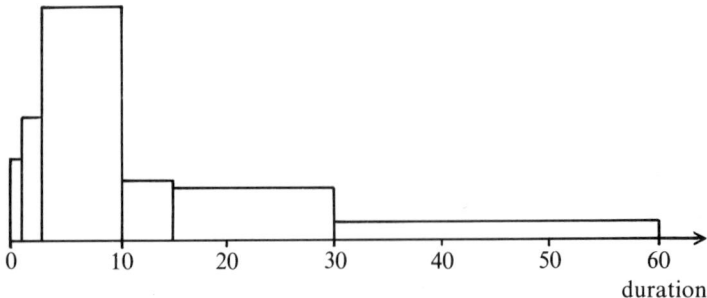

Comment critically on this illustration.

Draw your own histogram to represent these data and comment on any obviously different impressions it displays from those in the diagram above.

Find (i) the sample mean duration of the telephone calls,

(ii) the sample interquartile range of the durations of the telephone calls.

Telephone calls at peak rate are charged at a rate of 4.4p per minute. Provide an estimate of the mean cost of a call. (*JMB*)

50 **a** Find the maximum value of $x(1-x)$.

b Write down the formula for the standard error of a sample proportion in terms of the population proportion p and the sample size n. Show that, whatever the value of p, this standard error cannot exceed $\dfrac{1}{2\sqrt{n}}$.

c In a random sample of road accidents involving 92 seriously injured drivers, it was found that 69 of these drivers had not been wearing seat belts. Calculate an unbiased estimate for the population proportion p of seriously injured drivers not wearing seat belts and determine an approximate 95 per cent symmetric confidence interval for p.

d It is required to obtain a 95 per cent symmetric confidence interval for p. The sample size n is to be chosen such that the width of the confidence interval does not exceed 0.1.

(i) Use your estimate for p obtained in **c** to find, to the nearest 10, an approximate value for n.

(ii) Find the least value of n which is certain to achieve the required width of confidence interval whatever the value of p. (*JMB*)

51 Two workers, A and B, perform the same repetitive task in a factory. The times (in minutes) taken by A and B to complete individual tasks are independent and normally distributed with means μ_1 and μ_2, and standard deviations σ_1 and σ_2, respectively. During a time and motion study, records were kept of the time (x minutes) taken by each worker to complete each individual task. The times taken by A to complete 250 tasks were such that $\Sigma x = 7550$ and $\Sigma x^2 = 229\,006$, and the times taken by B to complete 200 tasks were such that $\Sigma x = 5950$ and $\Sigma x^2 = 178\,704$.

(i) Calculate unbiased estimates for $\mu_1, \mu_2, \sigma_1^2, \sigma_2^2$.

(ii) B claims that on average he completes a task in a shorter time than A. Stating suitable null and alternative hypotheses, determine whether there is sufficient evidence at the 5 per cent significance level to support B's claim.

(iii) Determine an approximate 95 per cent symmetric confidence interval for $\mu_1 - \mu_2$. State, with a reason, whether this discredits the assertion made by A that there is no difference in their mean rates of working. (*JMB*)

52 The relation between two quantities x and y is given by

$$y = \alpha + \beta x,$$

where α and β are unknown constants. Eight pairs of values of x and y were obtained experimentally and the following results calculated:

$$\Sigma x = 360, \quad \Sigma y = 704.8, \quad \Sigma x^2 = 20400, \quad \Sigma xy = 38142.$$

(i) Determine the least squares estimate of the equation $y = \alpha + \beta x$.

(ii) Assuming that the values of x are accurate and the observed values of y are subject to independent random errors, each normally distributed with mean zero and standard deviation 0.3, test, using a 5 per cent significance level, the null hypothesis $\beta = 1.52$ against the alternative hypothesis $\beta \neq 1.52$. State your conclusion about the size of β.

(JMB)

Answers

The Examining Boards listed in the Acknowledgements on page ii bear no responsibility whatever for the answers to examination questions given here, which are the sole responsibility of the author.

Chapter 1

Poisson distribution

Exercise 1.1

1	a	0.101	b	0.647	c	0.577		
2	a	0.130	b	0.849	c	0.970		
3	a	0.134	b	0.606	c	0.084		
4	a	0.139	b	0.140	c	0.937		
5	a	0.066	b	0.486	c	0.992		
6	a	0.607	b	0.090	c	0.007	d	0.265
7	a	0.125	b	0.542	c	0.029		
8	a	0.140	b	0.735	c	0.616		
9	a	0.301	b	0.361	c	2.96		
10	a	0.019	b	0.513	c	0.699		

Exercise 1.2

1	a	0.238	b	0.629; 0.232, 0.629	4	0.916		
2	0.785				5	a	0.342	b	0.468
3	0.036							

Exercise 1.3

1	a	0.067	b	0.523	c	0.532	4	a	0.276	b	0.276
2	a	0.637	b	0.452	c	0.829	5	a	0.196	b	0.484
3	a	0.179	b	0.092	c	0.081					

Miscellaneous Exercise 1

1 (i) 0.041 (ii) 0.019
2 (i) 0.05 (ii) 0.58; 0.07
3 (i) 0.300 (ii) 0.301; 0.34, 0.70
4 a (ii) $\frac{7}{15}$ b (i) 0.398 (ii) 0.0028
5 a $\frac{37}{256}$ b (i) 0.940 (ii) 0.038

STATISTICS 2

6 a (i) 0.058 (ii) 0.052 **b** 74 **c** 0.955

7 a (i) 0.68 (ii) 0.14 **b** 1.22 **c** £4·13

8 (i) 0.185 (ii) 4 (iii) 2.681 (iv) 6

9 (ii) 0.135, 0.271, 0.271, 0.323; 1.782 (iii) £7·69

10 0.018, 0.074, 0.146, 0.195, 0.196, 0.371; 13.59 (i) £3·94 (ii) not advisable

11 (i) 0.368 (ii) 0.053 (iii) 4

12 (i) 0.677 (ii) 0.543 (iii) 0.200

13 (i) 0.677 (ii) 0.017; 1497

14 0.371, 60.37

15 (i) 0.746 (ii) 18 (iii) 0.017

16 a (i) 0.135 (ii) 0.947 **b** (i) 0.144 (ii) 0.111 (iii) 0.046

17 a (i) 0.095 (ii) 0.090 **b** 0.853 **c** 0.632; 63.2, 4.82

18 (i) 2.22

(ii)

No. of calls	0	1	2	3	4	5	6	7
Rel. freq.	0.155	0.265	0.255	0.165	0.100	0.040	0.010	0.010
Poisson prob.	0.135	0.271	0.271	0.180	0.090	0.036	0.012	0.003

2; mean $\simeq$ variance, 0.195

19 (i) 0.938 (ii) 0.206 (iii) 46 (iv) 11

20 $r\mathrm{P}(Y = r) = \mathrm{P}(Y = r - 1)$ (i) 0.05 (ii) 0.16 (iii) 0.61

Chapter 2

The distribution of a function of a random variable

Exercise 2.1

1 $g(y) = \dfrac{1}{15}, 5 \leqslant y \leqslant 20$

2 $g(y) = \dfrac{y}{2}, 0 \leqslant y \leqslant 2$

3 $g(y) = \dfrac{(4 - y^{\frac{1}{2}})}{18}, 0 \leqslant y \leqslant 9$

4 $g(y) = \dfrac{1}{3}, -2 \leqslant y \leqslant 1$

5 $g(y) = 2\dfrac{(2y - 1)}{y^3}\cdot\dfrac{1}{2} \leqslant y \leqslant 1$

Exercise 2.2

1 $g(y) = 18\dfrac{(3y + 1)^2}{11}, \quad -\dfrac{1}{3} \leqslant y \leqslant 0$

$g(y) = 18\dfrac{(3y + 1)}{11}, \quad 0 < y \leqslant \dfrac{1}{3}$

2 $g(y) = \dfrac{y^{-\frac{3}{2}}}{18}, \quad \dfrac{1}{36} \leqslant y < \dfrac{1}{9}$

$g(y) = \dfrac{y^{-\frac{1}{2}}}{2}, \quad \dfrac{1}{9} \leqslant y \leqslant 1$

3 $g(y) = 1, \quad 0 \leqslant y \leqslant 1$

4 $g(y) = \dfrac{(6 - 2y^2)}{9}, \quad 0 \leqslant y \leqslant 1$

$g(y) = \dfrac{(3 + 2y - y^2)}{9}, \quad 1 < y \leqslant 2$

5 $g(y) = 2\dfrac{(5 - y)}{9}, \quad 2 \leqslant y \leqslant 5$

Miscellaneous Exercise $\boxed{2}$

1 $g(y) = \dfrac{5y^{-\frac{1}{2}}}{2}$, $\quad 98.01 \leqslant y \leqslant 102.01$

2 $\dfrac{(b+a)}{2}, \dfrac{(b-a)^2}{12}$; $\quad g(y) = 1, 0 \leqslant y \leqslant 1$; $\quad 0.95$; $\quad -1, 1$

3 (i) 24.3 $\qquad$ (ii) $F(x) = 0, x < 20$; $\quad F(x) = 3 - \dfrac{60}{x}, 20 \leqslant x \leqslant 30$; $\quad F(x) = 1, x > 30$

$\qquad$ (iii) 24 $\qquad$ (iv) $g(y) = \dfrac{1}{5}, 10 \leqslant y \leqslant 15$

4 $f(x) = \dfrac{2}{\pi}, 0 \leqslant x \leqslant \dfrac{\pi}{2}$; $\quad F(x) = \dfrac{2x}{\pi}, 0 \leqslant x \leqslant \dfrac{\pi}{2}$

$\qquad g(y) = \dfrac{2}{[\pi\sqrt{(4r^2 - y^2)}]}, 0 \leqslant y \leqslant 2r$

5 $-1, 1$; $\quad 0$

6 (i) $\dfrac{1}{8}$ $\qquad$ (ii) $\dfrac{5}{16}, \dfrac{21}{16}$ $\qquad$ (iii) $g(y) = 2, \dfrac{17}{16} \leqslant y \leqslant \dfrac{25}{16}$

7 (i) $\dfrac{3}{4}$ $\qquad$ (iii) $F(x) = \dfrac{3x^2}{4} - \dfrac{x^3}{4}, 0 \leqslant x \leqslant 2$ $\qquad$ (iv) $R = [0, 1], k = 24$

8 (i) $\dfrac{13}{12}$ $\qquad$ (ii) $F(x) = \dfrac{x}{2}, 0 < x < 1$; $\quad F(x) = 1 - \dfrac{(3-x)^2}{8}, 1 \leqslant x \leqslant 3$;

$\qquad$ (iii) $\dfrac{5}{3}$ $\qquad$ (iv) $g(y) = \dfrac{y}{4}, 0 \leqslant y \leqslant 2$; $\quad g(y) = \dfrac{1}{2}, 2 < y \leqslant 3$

9 (i) $\dfrac{5}{3}, \dfrac{2}{3}$ $\qquad$ (ii) $\dfrac{2}{3}$ $\qquad$ (iii) $G(y) = \dfrac{25y}{24} - \dfrac{1}{24}, \dfrac{1}{25} \leqslant y \leqslant 1$

$\qquad\qquad\qquad\qquad\qquad\qquad g(y) = \dfrac{25}{24}, \qquad \dfrac{1}{25} \leqslant y \leqslant 1$

10 (i) $F(x) = \dfrac{x}{8}, 0 < x < 2$; $F(x) = \dfrac{x^2}{16}, 2 \leqslant x \leqslant 4$ $\qquad$ (ii) $(a) \dfrac{7}{16}, (b) \dfrac{7}{9}$

$\qquad$ (iii) $\quad g(y) = \dfrac{3}{4}, 0 \leqslant y \leqslant 1$; $\quad \dfrac{7}{8}$

11 $h(w) = 2(3 - w), 2 \leqslant w \leqslant 3$; $\quad j(y) = 4(y - 2)^3, 2 \leqslant y \leqslant 3$

12 (i) $F(x) = \dfrac{2x}{5}, 0 < x < 1$; $\quad F(x) = 1 - \dfrac{(4-x)^2}{15}, 1 \leqslant x \leqslant 4$; $\quad 0.625, 1.26$

$\qquad$ (ii) $G(y) = \dfrac{9}{5} - \dfrac{24}{5y}, 4 < y < 6$;

$\qquad\qquad g(y) = \dfrac{48}{5y^2}\left(1 - \dfrac{2}{y}\right), 2 \leqslant y \leqslant 4$; $\quad g(y) = \dfrac{24}{5y^2}, 4 < y < 6$

STATISTICS 2

Chapter 3

Joint distributions of discrete random variables

Exercise 3.1

1 a

			x	
		0	1	2
	0	p	$8p$	$6p$
y	1	$6p$	$12p$	0
	2	$3p$	0	0

where $p = \dfrac{1}{36}$

b $\dfrac{8}{9}, \dfrac{35}{81}; \dfrac{2}{3}, \dfrac{7}{18}; \dfrac{14}{9}, \dfrac{49}{162}$

2 a

				x	
		0	1	2	3
	0	p	0	0	0
y	1	0	$3p$	p	0
	2	0	0	$2p$	0
	3	0	0	0	p

where $p = \dfrac{1}{8}$

b $\dfrac{3}{2}, \dfrac{11}{8}, \dfrac{11}{4}$

3 a

					x			
		0	1	2	3	4	5	6
	0	0	0	0	$8p$	$12p$	$6p$	p
y	1	0	0	$36p$	$36p$	$9p$	0	0
	2	0	$54p$	$27p$	0	0	0	0
	3	$27p$	0	0	0	0	0	0

where $p = \dfrac{1}{216}$

b $2, \dfrac{3}{2}, \dfrac{7}{2}$

Exercise 3.2

1 a $\dfrac{1}{20}$ **c** $\dfrac{23}{10}, \dfrac{13}{10}$

d

z_i	0	1	2	3	4	6
p_i	$4p$	p	$4p$	$3p$	$3p$	$5p$

2 a $\dfrac{1}{20}$ **c** $1, \dfrac{3}{5}$

d

z_i	0	1	2
p_i	$11p$	$6p$	$3p$

3 a $\dfrac{1}{12}$ **c** $1, 1$

d

z_i	0	2	4
p_i	$8p$	$2p$	$2p$

214

Exercise $\boxed{3.3}$

1 **a** 16, 7.2 **b** 8, 7.2 **c** 44, 52.8 **4** **a** 0.100 **b** 0.587

2 **a** 10, 10 **b** 2, 10 **c** 26, 70 **5** **a** 0.971 **b** 2, 34

3 0.185

Miscellaneous Exercise $\boxed{3}$

1 (i) X and Y are not independent

2 (i) B(3, 0.4), 1.2, 0.72; B(3, 0.2), 0.6, 0.48

(ii) 0.192, 0.064, 0.192, 0.048, 0.008 (iii) 0.352

3 (ii) 2.45 (iii)

		u	
		2	4
w	0	0.05	0.15
	2	0.2	0.6

$Y - 3X = 2W - U$; 3.31

4

		x		
		0	1	2
y	0	$5p$	$10p$	$3p$
	1	$5p$	$6p$	$1p$

where $p = \dfrac{1}{30}$

z_i	p_i
0	$5p$
1	$15p$
2	$9p$
3	p

$E[Z] = 1.2$
$V[Z] = 0.56$

B(4, 0.3), 1.2, 0.84

5

		x			
		0	1	2	3
	0	$4p$	0	0	0
y	1	0	$2p$	$5p$	$9p$
	2	0	$2p$	$6p$	0
	3	0	$4p$	0	0

where $p = \dfrac{1}{32}$

$\dfrac{5}{4}, \dfrac{3}{4}$

6 (i) 0.191 (ii) 0.018 (iii) 0.264; 2

7 **a** (i) 0.986 (ii) 0.223 **c** 6p, 6p

8 **a** (i) $e^{-(\lambda + \mu)}$ (ii) $\dfrac{\lambda^2}{2} \mu e^{-(\lambda + \mu)}$ (iii) $e^{-(\lambda + \mu)} \dfrac{(\lambda + \mu)^3}{3!}$

b (i) 0.001 (ii) 0.140 (iii) $\dfrac{135}{512}$

9 0.392, 0.0062, £1·74, 50p

10 (i) $\dfrac{1}{4} e^{-1}$ (ii) $\dfrac{7}{8} e^{-1}$ (iii) $\dfrac{1}{4}(1 + 3e^{-1})$ (iv) 1, 2

Chapter 4

Joint distributions of continuous random variables

Exercise 4.1

1 a 0.048   b 0.313   4 0.841
2 0.836   5 10.44 a.m.
3 a 530, 12, C and E independent   b 0.0062

Miscellaneous Exercise 4

1 (i) 0.159   (ii) 0.191   (iii) 0.584   6 0.0625, 0.2574;   0.5, 0.7123;   7
2 (i) 0.945   (ii) 0.975   7 (i) 0.106   (ii) 0.266
3 0.8413;   N(3, 2.5), 0.9711   8 1000, 172; 3μ, $\sqrt{3}\sigma$;   0.16;   0.02
4 (i) 0.023   (ii) 0.683;   0, 24   9 (ii) 189   (iii) 0.430   (iv) 0.023
5 (i) 0.4602 (ii) 248.9 (iii) 0.5793 (iv) 24.3g, 1.5g   10 (i) 0.425   (ii) 0.153

Chapter 5

Sampling distributions

Exercise 5.1

1

x_i	4	$\frac{14}{3}$	$\frac{16}{3}$	6
p_i	0.1	0.3	0.3	0.3

5.2, 0.653

2 3;

m_i	2	3	4
p_i	0.3	0.4	0.3

unbiased

3

x_i	2	3
p_i	$\frac{2}{5}$	$\frac{3}{5}$

4 2, 2;

v_i	0.25	1	2.25	4
p_i	0.4	0.3	0.2	0.1

5

y_i	0	0.5	1
p_i	0.1	0.6	0.3

Exercise 5.2

1

x_i	4	$\frac{14}{3}$	$\frac{16}{3}$	6	$\frac{20}{3}$	$\frac{22}{3}$	8
p_i	$27p$	$27p$	$36p$	$19p$	$12p$	$3p$	p

5.2, 0.924

where $p = \dfrac{1}{125}$

2

m_i	1	2	3	4	5
p_i	$13p$	$31p$	$37p$	$31p$	$13p$

3

where $p = \dfrac{1}{125}$

3 3, 2;

v_i	0	0.25	1	2.25	4
p_i	$5p$	$8p$	$6p$	$4p$	$2p$

where $p = \dfrac{1}{25}$

4 1.6, 0.4

5 $\dfrac{2}{3}, \dfrac{1}{18}$; $\dfrac{2}{3}, 0.1054$

Exercise 5.3

1 **a** 0.0304   **b** 0.8882   **4** 68
2 **a** 0.6795   **b** 0.9992   **5** 0.7881   **a** 0.386   **b** 0.9452
3 385

Exercise 5.4

1 0.0339   **4** 0.9969
2 0.0386
3 0.9599   **5** $\dfrac{2}{3}, \dfrac{1}{18}$; 0.9772

Miscellaneous Exercise 5

1 3.6, 5.04; 0.84; 1.68

2 $2, \dfrac{2}{3}$;

$\bar{x}_i$	1	$\dfrac{4}{3}$	$\dfrac{5}{3}$	2	$\dfrac{7}{3}$	$\dfrac{8}{3}$	3
p_i	$\dfrac{1}{27}$	$\dfrac{3}{27}$	$\dfrac{6}{27}$	$\dfrac{7}{27}$	$\dfrac{6}{27}$	$\dfrac{3}{27}$	$\dfrac{1}{27}$

m_i	1	2	3
p_i	$\dfrac{7}{27}$	$\dfrac{13}{27}$	$\dfrac{7}{27}$

(i) $\dfrac{2}{9}$ (ii) $\dfrac{14}{27}$ (iii) $\bar{X}$, smaller variance

3 **a** 4, 14.4;

 b (i)

x_i	$\dfrac{4}{3}$	2	$\dfrac{10}{3}$	$\dfrac{14}{3}$	$\dfrac{16}{3}$	$\dfrac{20}{3}$
p_i	0.1	0.1	0.3	0.2	0.2	0.1

 (ii) unbiased   (iii) 2.4

4 2.5, 1.25; (i) 2.5, $\dfrac{5}{12}$;   (ii) $\dfrac{5}{32}, \dfrac{11}{32}, \dfrac{11}{32}$; $\dfrac{7}{8}$

5 19.8, 0.6;

m_i	19	$\dfrac{58}{3}$	$\dfrac{59}{3}$	20	$\dfrac{61}{3}$
p_i	p	$18p$	$48p$	$38p$	$15p$

where $p = \dfrac{1}{120}$

(ii) 0.3055

6 (ii) $\dfrac{\pi}{2}$ (iv) 0.967

9 (i) $\dfrac{13}{12}, \dfrac{71}{144}$ (ii) $\dfrac{15}{4}$ (iii) 0.118

7 (i) 0.5 (ii) 0.159

10 (i) 1.2, 1.36 (ii) 1.2, 0.023

8 0.8664

11 (i) 0.893 (ii) 5380 (iii) 0.994

12 N(54, 64), 0.0228; 15.68; N(5, 144), 0.6616

13 a (i) 0.7881 (ii) 0.6730 b 0.0746 c 0.0548

14 (i) 145 (ii) 0.8413 (iv) 0.0062

15 $\dfrac{1}{12}$; $0, \dfrac{n}{12}$; 0.9168; 16

Chapter 6

Estimation

Exercise 6.1

1 T_1, T_2, T_4

4 a $(a+b)\mu, (a^2+b^2)\sigma^2$ b $a = \dfrac{1}{2}, b = \dfrac{1}{2}$

2 T_4

5 $a = \dfrac{1}{3}, b = \dfrac{1}{3}$

3 T_1

Exercise 6.2

1 0.35, 0.0151, 0.0158

4 0.84, 0.75, 0.804 (smallest SE)

2 0.24, 0.0604, 0.0707

5 $\hat{P}_2$

3 0.92, 0.0192, 0.0354

Exercise 6.3

1 2.7, 0.1429

4 14.64, 3.199

2 35.3, 1.652

5 2.111, 2.158

3 176.25, 103.5

Miscellaneous Exercise 6

1 (i) 8 (ii) 0.777, 0.013

3 60%

5 5, 4.5; 24

2 5.9, 8.1

4 T_1 better

6 (i) $\dfrac{(2+\theta)}{4}$ (ii) $B\left(n, \dfrac{(2+\theta)}{4}\right)$ (iii) T_2 better, since $V[T_2] < V[T_1]$

7 $3 - 3\theta, 5\theta - 9\theta^2$; P_2 has the smaller standard error

8

t_i	0	1	4
p_i	$1 - 2\theta + \dfrac{3\theta^2}{2}$	$2\theta - 2\theta^2$	$\dfrac{\theta^2}{2}$

$V[Y] = \dfrac{(\theta + \theta^2)}{2}$

Z has the smaller variance

9 Z_2 has the smaller variance

10 $a, \dfrac{c^2}{3}$; $\dfrac{c^2(6a^2 + c^2)}{9}$

Chapter 7

Confidence intervals
Exercise 7.1
1 a (160.8, 163.2) **b** (161.0, 163.0) **c** (160.5, 163.5) **4 a** (964.9, 1001.1)
2 a (77.6, 87.4) **b** (78.4, 86.6) **c** (76.1, 88.9) **b** 290
3 a (1.58, 1.62) **b** 385 **5** 148, 49

Exercise 7.2
1 441.96, 2.2509, (441.71, 442.21) **4** (979.235, 989.105), (12.33, 29.53)
2 17.64, (80.60, 82.20) **5** (1.572, 1.588), (0.126, 0.156)
3 (16.3, 25.7)

Exercise 7.3
1 7, 4, (6.608, 7.392) **3** (9.13, 11.43)
2 (70.85, 71.55) **4** (131.62, 132.38), (2.38, 3.62)
5 (17.69, 18.11), (−0.05, 0.45), no, since 0 is contained in the interval

Exercise 7.4
1 (0.396, 0.458)
2 a (0.363, 0.477) **b** 6592 **c** 6766
3 a (0.059, 0.221) **b** (148, 552)
4 a 8000 **b** (0.0275, 0.0725) **c** (5500, 14500)
5 a (0.113, 0.187) **b** (0.078, 0.164) **c** (−0.028, 0.086)

Exercise 7.5
1 (7.09, 7.99) **4** (13.02, 15.05)
2 (3.38, 4.82), sample was random (doubtful) **5** (0.25, 0.87)
3 (12.98, 15.02)

Exercise 7.6
1 (70.18, 71.82)
2 (186.6, 190.6), heights normally distributed, random sample chosen
3 (1.31, 4.89), interval supports the claim as it contains 3.5
4 (43.90, 46.10) **5** 12, 4

Miscellaneous Exercise 7
1 (0.487, 0.533), truthful **3** (3.841, 3.907)
2 2500, (0.167, 0.233), (2100, 3000) **4** 6, 4, (5.38, 6.62)
5 0.065, 0.0174, (0.031, 0.099), supports, 0.064 (Poisson), 0.052 (normal)
6 a (i) 9.875, 0.25 (ii) (9.765, 9.985) **b** 385

7 $0.64, 0.048, (0.55, 0.73), \theta = \dfrac{(\alpha - 2)}{2}, (3.09, 3.47)$

8 (i)(6.42, 8.38)   (ii)62

9 $\theta + \dfrac{3}{2}, \dfrac{1}{12}$;   1.9, 0.4, 0.0913;   (0.303, 0.417)

10 $\bar{X}$ approximately $N\left(\mu_1, \dfrac{\sigma_1^2}{n_1}\right)$ $\bar{Y} - \bar{X}$ approximately $N\left(\mu_2 - \mu_1, \dfrac{\sigma_1^2}{n_1} + \dfrac{\sigma_2^2}{n_2}\right)$

(1.55, 1.65), (0.102, 0.298)

11 (i)28.72   (ii)0.3785;   (28.55, 28.89)

12 0.51, 1.49, (1.745, 5.098)

13 (i)385   (ii)$0.804\lambda, 1.196\lambda$;   (46, 68)

14 (18.02, 19.98)

15 $f(x) = \dfrac{1}{(a - 10)}, 10 \leqslant x \leqslant a$; $\dfrac{(a + 10)}{2}$;   32;   34, (25.8, 42.2)

16 (i) $1.7, \dfrac{1}{7}$ (ii)(1.38, 2.02)

17 (i)(14.24, 16.36)   (ii)0.9876

18 a 0.8664   **b** (i)2   (ii)(49.0, 51.0)

19 a (0.261, 1.439); weight losses normally distributed, sample random; interval supports claim as it contains 1
  b (0.290, 1.030); weight losses are independent normally distributed random variables, samples are both random; diet A is superior to diet B with 95% confidence

20 a (339.7, 342.3), consistent, because 340 is contained in the interval
  b (i)(6.93, 7.27)   (ii)(1.14, 1.46)

21 a (i) $-0.02, 0.012$ (ii)$(-0.124, 0.084)$
  b (85.31, 87.49); $(-0.12, 2.52)$; no, because 0 is contained in the interval

22 (i)456, 108   (ii)(449.4, 462.6)   (iii)457, 89.18

23 a $(-0.25, 0.73)$ **b** 139

24 (i)normal   (ii)none;   (13.36, 15.84), (13.03, 16.17)

25 $\bar{X}$ approximately $N\left(\mu, \dfrac{\sigma^2}{n}\right)$ (i)15.2   (ii)8.9984   (iii)(13.99, 16.41)
                                                                        8.6857   (14.84, 15.56)

Chapter 8

Hypothesis testing
Exercise 8.1

1 $H_0: p = 0.55$, $H_1: p > 0.55$; {8, 9, 10}; result is significant at 10% level. Conclude that the modified drug is an improvement.

2 1%; result is significant at 1% level and conclude that $p > 0.4$.

3 $H_0: p = 0.7$, $H_1: p > 0.7$; 3.5%

4 $H_0: p = 0.2$, $H_1: p > 0.2$; {8, 9, 10, ..., 20}; 3.2%

5 $H_0: p = 0.5$, $H_1: p \neq 0.5$; {0, 1, 2, 10, 11, 12} with $\alpha = 3.9\%$

Exercise 8.2

1 $H_0: p = 0.36$, $H_1: p > 0.36$; $z = 1.82 > 1.64$ result is significant at 5% level. Accept H_1 and conclude that the campaign was successful.

2 $H_0: p = 0.5$, $H_1: p \neq 0.5$; at 5% level
   a $z = 2.19 > 1.96$, result significant. Coin biased towards heads.
   b $z = 1.34 < 1.96$, result not significant. No reason to reject H_0.
   c $z = -2.05 < -1.96$, result significant. Coin biased towards tails.

3 $H_0: p = 0.75$, $H_1: p < 0.75$; at 10% level
   a $z = -1.04 > -1.28$, result not significant. Do not reject H_0.
   b $z = -1.55 < -1.28$, result significant. Reject the company's claim.

4 $H_0: p = 0.55$, $H_1: p \neq 0.55$; $z = -2.63 < -2.58$, result significant at 1% level. Conclude that the proportion of males < 0.55 for wild rabbits.

5 $H_0: p = 0.8$, $H_1: p > 0.8$; $z = 2.40 > 2.33$, result significant at 1% level. Conclude that the treatment has increased the germination rate.

6 $H_0: p = \dfrac{1}{37}$, $H_1: p > \dfrac{1}{37}$; $z = 3.29 > 3.09$, result significant at 0.1% level. Conclude that the wheel is unfair.

Exercise 8.3

1 $H_0: \mu = 51.62$, $H_1: \mu < 51.62$; $z = -1.86 < -1.64$, result is significant at 5% level. Conclude that altitude training (or some other factor) has improved the runner's average performance.

2 $H_0: \mu = 500$, $H_1: \mu \neq 500$; $z = 1.75 < 1.96$, result not significant at 5% level. Not enough evidence to suggest that the machine needs adjustment.

3 $H_0: \mu = 41.2$, $H_1: \mu > 41.2$; $z = 1.96 > 1.28$, result significant at 10% level. Conclude that the modification has improved the fuel consumption.

4 $H_0: \mu = 5.4$, $H_1: \mu \neq 5.4$; $z = 1.79 > 1.64$, result significant at 10% level. Conclude that A is not the author of the fragment.

5 $H_0: \mu = 454$, $H_1: \mu < 454$; $z = -1.47 > -2.33$, result not significant at 1% level. Not enough evidence to support the claim that the line is producing articles with a mean below the nominal mass.

Exercise 8.4

1 $H_0: \mu = 120$, $H_1: \mu < 120$; $t = -2.35 < -1.89$, result significant at 5% level. Foreman's suspicion confirmed. Assumptions: number of articles produced per hour is normally distributed and that the eight numbers given in the question form a random sample.

2 $H_0: \mu = 0$, $H_1: \mu \neq 0$; $t = 2.32 > 2.23$, result significant at 5% level. Conclude that the drug has increased the mean blood pressure of patients. Assumption: the changes in blood pressure are normally distributed.

3 $H_0: \mu = 24$, $H_1: \mu \neq 24$; $t = -1.94 < -1.76$, result significant at 10% level. Conclude that the mean lifetime of the batteries is less than 24 h.

4 $H_0: \mu = 1.5$, $H_1: \mu < 1.5$; $t = -1.53 > -1.72$, result not significant at 5% level. The packs are not significantly underweight.

5 $H_0: \mu = 1000$, $H_1: \mu < 1000$; $t = -2.52 > -3.75$, result not significant at 1% level. Do not reject H_0.

Exercise 8.5

1 $H_0: \mu_1 - \mu_2 = 0$, $H_1: \mu_1 - \mu_2 \neq 0$; $z = 1.76 < 1.96$, result not significant at 5% level. The performances of the two packers are not significantly different.

2 $H_0: \mu_1 - \mu_2 = 0$, $H_1: \mu_1 - \mu_2 > 0$; $z = 0.99 < 1.64$, result not significant at 5% level. Do not reject H_0. Reject the company's claim.

3 $H_0: \mu_2 - \mu_1 = 2$, $H_1: \mu_2 - \mu_1 > 2$; $z = 2.48 > 2.33$, result significant at 1% level. Accept H_1.

4 $H_0: \mu_B - \mu_A = 5$, $H_1: \mu_B - \mu_A > 5$; $z = 1.44 > 1.28$, result significant at 10% level. Accept H_1.

5 $H_0: \mu_1 - \mu_2 = 0$, $H_1: \mu_1 - \mu_2 \neq 0$; $z = 1.98 > 1.64$, result significant at 10% level. Conclude that the workers using Method B are quicker on average than those using Method A.

Exercise 8.6

1 $H_0: \alpha = 1.25$, $H_1: \alpha > 1.25$; $P(Y \geqslant 10) = 0.032 < 0.05$, result significant at 5% level. Conclude that the mean number of claims per week has increased.

2 $H_0: \alpha = 1.4$, $H_1: \alpha < 1.4$; $P(Y \leqslant 3) = 0.082 < 0.10$, result significant at 10% level. Conclude that the training has reduced the mean number of errors per page.

3 $H_0: \alpha = 5$, $H_1: \alpha > 5$; $z = 1.19 < 1.64$, result not significant at 5% level. Not enough evidence to suggest that the TV campaign has been successful.

4 $H_0: \alpha = 0.2$, $H_1: \alpha > 0.2$; $P(Y \geqslant 6) = 0.017 > 0.01$, result not significant at 1% level. Conclude that the roll is acceptable.

5 $H_0: \alpha = 4$, $H_1: \alpha \neq 4$; $z = -1.75 > -1.96$, result not significant at 5% level. Not enough evidence to support the claim that the mean number of accidents has changed.

Exercise 8.7

1 $(21.312, \infty)$; 0.983

2 $(-\infty, 41.648) \cup (46.352, \infty)$; 0.133

3 0.068, 0.593

4 0.090, 0.011

5 0.058, 0.250

Miscellaneous Exercise 8

1 $H_0: a = 10$, $H_1: a \neq 10$; $P(X \leqslant 4) = 0.029 > 0.025$, result not significant at 5% level, retain H_0.

2 (i) (4.76, 5.74) (ii) (0.230, 0.595) $[\Phi(5 - 4.76) = 0.595, \Phi(5 - 5.74) = 0.230]$
$z = 1.80 < 1.96$, result not significant at 5% level, retain H_0.

3 $z = 2.37 > 2.33$, result significant at 1% level, conclude $\mu > 17.5$.

4 12.57, 1.89; 13.64, 3.20; $H_0: \mu_B - \mu_A = 1$, $H_1: \mu_B - \mu_A > 1$, $z = 0.47 < 1.28$, result is not significant at 10% level, retain H_0.

5 15.2, 0.09; (15.15, 15.25); Central Limit Theorem; $H_0: \mu = \mu_1$, $H_1: \mu < \mu_1$, $z = 1.735 > 1.64$, result significant at 5% level, conclude $\mu < \mu_1$.

6 251.9, 36; (251.41, 252.39); $H_0: \mu_A = \mu_B$, $H_1: \mu_A \neq \mu_B$, $z = 1.8 < 1.96$, result not significant at 5% level, retain H_0. $H_0: \mu_A = \mu_B$, $H_1: \mu_A > \mu_B$, $z = 1.8 > 1.64$, result significant at 5% level, reject H_0 in favour of H_1.

7 (i) 0.335 (ii) 5 (iii) 4 (iv) $4e^{-3}$ (v) $P(X \leqslant 4 | \alpha = 9.2) = 0.049 < 0.05$, result significant at 5% level, conclude safety precautions effective.

8 52.5, 2.806; (52.04, 52.96);
(i) $z = 2.11 > 1.96$, result significant at 5% level, conclude $\mu > 52$,
(ii) $z = 2.11 < 2.58$, result not significant at 2% level, retain $H_0: \mu = 52$,
$z = 1.00 < 1.96$, result not significant at 5% level, conclude means equal.

9 **a** (i) 0.0228   (ii) 0.309   **b** $c = 6 - \dfrac{2.632}{\sqrt{n}}$; $n = 33$, $c = 5.54$.

10 (i) $(-\infty, 70.236) \cup (73.764, \infty)$ (ii) 0.085, 0.915

11 0.042, 0.383

12 $H_0: p = \dfrac{1}{2}, H_1: p > \dfrac{1}{2}$; 0.021, 0.584

13 0.821; $P(C \geqslant 8) = 0.055 > 0.05$, result not significant at 5% level, retain H_0.
$z = 2.4 > 1.64$, result significant at 5% level, conclude E superior to F.

14 (i) $P(X \leqslant 2) = 0.055 > 0.05$, result not significant at 10% level, coin unbiased,
(ii) $z = 1.7 > 1.64$, result significant at 10% level, conclude coin is biased towards heads.

15 $H_0: \alpha = 7.5$, $H_1: \alpha < 7.5$;
(i) $P(X \leqslant 3) = 0.059 > 0.05$, result not significant at 5% level, ineffective,
(ii) $z = -2.10 < -1.64$, result significant at 5% level, spray effective.

16 $N\left(\mu_1 - \mu_2, \dfrac{\sigma_1{}^2}{m} + \dfrac{\sigma_2{}^2}{n}\right)$; (0.032, 1.068)
(i) $z = 0.29 < 1.96$, result not significant at 5% level, retain $H_0: \mu_1 = 3$,
(ii) $z = 2.08 > 1.96$, result significant at 5% level, conclude $\mu_1 < \mu_2$.

17 $\dfrac{\theta}{(2 + \theta)}$; P(no. of negative values $\geqslant 17$) = 0.016 < 0.02, result is significant at 2%
level, conclude $\theta > 3$.

18 34

19 (i) $F(x) = 1 - x^{-\alpha}$, $x \geqslant 1$; median = $\sqrt{2}$;
(ii) R is approximately Po(4), $P(R \geqslant 9) = 0.021 < 0.025$, result significant at 5%
level, conclude $\alpha < 2$.

20 (i) $c = 1.849$

<hr>

Chapter 9

Linear relationships

Exercise 9.1

1 **a** 2.44   **b** 6.1   **3** **a** $y = -4 + 0.8x$ **b** 12
2 1.7, 2.55   **4** **a** $y = 2 + 2x$ **b** 22
5 $y = -1 + 4x$; the relation between x and y is $y = x^2 + 1$, which is non-linear

Exercise 9.2

1 **a** $y = 1.8 + 2.4x$ **b** (2.385, 2.415)   **c** (1.721, 1.879)   **d** (19.73, 19.87)
2 **a** $y = 15.6 - 0.8x$ **b** (−0.849, −0.751)   **c** (15.07, 16.13)   **d** (7.32, 7.88)
3 **a** 62.5   **b** $z = -7.45 < -1.96$, significant at 5% level, conclude $y_0 < 63$
  c (1.191, 1.209)   **d** (23.82, 24.18)
4 **a** 3.2, 3.9
  b $z = -1.58 > -1.64$, not sufficient evidence at 5% level to support the hypothesis $f < 4$
5 **a** (1.592, 1.628)   **b** (7.96, 8.14)

STATISTICS 2

Exercise 9.3

1 10.16, −1.01   **3** 18.7   **5** 770
2 7.72, 108   **4** 1.99, 7350

Exercise 9.4

1 a $y = 30.9 - 1.6x$ **b** (14.70, 15.10)   **c** (14.72, 15.48)
2 a $y = 13.46 + 0.329x$
  b 46.36, since 100 dB is outside the range of given data this estimate is unreliable.
3 a (54.79, 55.41)   **b** (55.43, 56.57)   **c** the one in **a**
4 a 21.8   **b** (19.91, 20.09)   **5** 1.728 (x on y)

Miscellaneous Exercise 9

1 (i) $y = 3.7 + 4.2x$ (ii) (3.23, 4.17)
2 (i) 10.4   (ii) (13.33, 16.67)   (iii) (5.18, 8.62)
3 (i) $y = 16 + 0.4x$ (ii) (0.33, 0.47), (6.6, 9.4)
4 a $y = 36 + 13x$ **b** (i) (10.52, 15.48)   (ii) (78.62, 84.38)
5 (i) $y = -71.2 + 3.5x$ (ii) 138.8 g   (iii) (138.03, 139.57)
6 (i) 5.11, 1.67   (ii) $z = 1.74 > 1.64$, significant at 5% level, conclude $f > 5$
7 (i) $y = 24 - 4x$
  (ii) 8, SE$[\hat{Y}_0] = 0.3953$; $\hat{Y}_0$ preferable to $\bar{Y}_4$ because SE$[\hat{Y}_0] < 0.7217 =$ SE$[\bar{Y}_4]$
  (iii) (7.35, 8.65)
8 (i) 16.25   (ii) (a) (12.46, 14.04)   (b) (1.00, 1.40)
9 (i) 38.8, 12.4   (ii) (37.94, 39.66)   (iii) (1.06, 1.94)
10 (i) 2.52   (ii) (2.47, 2.57), (7.40, 7.72)

Chapter 10

Revision

Exercise 10.1

1 (i) 0.584   (ii) 0.630   (iii) 0.244   (iv) 0.577   **2** (i) 0.315   (ii) 0.583
3 (i)(a) 0.146   (b) 0.785   (ii) 0.213   (iii) 3.59
4 (i) 0.199   (ii) 0.185   (iii) $e^{-3}(1 - e^{-3})0.04731$ (iv) $3e^{-6}(1 - e^{-3})0.007066$; 0.870
5 (i) 3   (ii) 0.103   (iii) 0.229   (iv) 0.090   (v) £1·95

Exercise 10.2

1 $g(y) = \dfrac{y^{\frac{1}{8}}}{8}, 0 \leqslant y \leqslant 16$ **2** $g(y) = \dfrac{(7 - y)}{8}, 2 \leqslant y \leqslant 4$

3 $g(y) = \dfrac{1}{40}, 40 \leqslant y \leqslant 80$

4 a $\dfrac{a^2}{18}$ **b** $F(x) = \dfrac{x^2}{a^2}, 0 \leqslant x \leqslant a$ **c** (i) $(a + 1)^{-2}$ (ii) $2(1 - y), 0 \leqslant y \leqslant 1$
5 (i) $1 - x^{-3}$ (ii) 1.5, 0.75; $g(y) = 3y^2, 0 \leqslant y \leqslant 1$; X and Y not independent

224

Exercise 10.3

1 **a** 13, 20 **b** 2, 18 **c** 8, 95 **3** (i) $\dfrac{1}{64}$, 8, 12, 6, 1 (iii) $\dfrac{9}{4}, \dfrac{3}{4}$

2 (ii) 0.7576 (iii) 13, 8 **4** (ii) 1, 2 (iii) not independent

(iv) $4\alpha + 1$

5 $1 - p$, $p(1 - p)$;

(i)

		x	
		0	1
y	0	p^2	$p(1 - p)$
	1	$p(1 - p)$	$(1 - p)^2$

$E[XY] = E[X].E[Y]$ for all p

(ii)

		x	
		0	1
y	0	p^2	$(1 - p)^2$
	1	$p(1 - p)$	$p(1 - p)$

$E[XY] = E[X].E[Y]$ when $p = \dfrac{1}{2}$

Exercise 10.4

1 24, 15 **2** (i) 11, 4 (ii) 64, 14

3 (i) 0.0139 (ii) 0.1587 (iii) 0.9332

4 (i) 0.006 (ii) 3.733 (iii) 0.910 (iv) 0.336 (v) 0.110

5 (i) 12.4, 0.3 (iii) 0.56 (iv) 3 phases of the triple jump are unlikely to be independent

Exercise 10.5

1 (i) $\dfrac{2}{11}$ (ii) $\dfrac{5}{11}$; $\dfrac{18}{11}, \dfrac{72}{121}$; $\dfrac{18}{11}, \dfrac{90}{121}$; 13

2 (i) 2, 0.2 (ii) 0.944, 0.028; 2, 0.056 (iii) 2, $\dfrac{1}{15}$

3 (i) $\dfrac{15}{64}$ (ii) 0.6915 (iii) 10.00 a.m.

4 (i) 0.0228 (ii) 0.7341; 162.5, 0.25; 173.5, 0.64; 11, 0.89; 0.1446; no, bimodal

5 (i) 0.081 (ii) 0.168; 0.098

Exercise 10.6

1 0.8

t_i	0	0.5	2
p_i	0.36	0.32	0.32

2 **a** (ii)

x_i	0	$\dfrac{1}{4}$	$\dfrac{1}{2}$	$\dfrac{3}{4}$
p_i	$\dfrac{1}{6}$	$\dfrac{1}{2}$	$\dfrac{3}{10}$	$\dfrac{1}{30}$

(iv) 0.187

b 0.3, 0.229

c X, since $SE[X] < SE[Y]$

3 $\theta + 1.5, \dfrac{1}{12}$; 1.9, 0.4, 0.0913; (0.303, 0.417)

4 $\dfrac{y^n}{\theta^n}; \dfrac{\theta^2}{n(n+2)}$

5

x_i	1	1.5	2	2.5	3
p_i	p	$18p$	$26p$	$18p$	p

where $p = \dfrac{1}{64}$; 33:32

Exercise 10.7

1 50.25, 1.907; (49.89, 50.61); (0.24, 1.06)

2 175.25, 103.5; (174.07, 176.43); Central Limit Theorem; $H_0: \mu_M - \mu_W = 10$, $H_1: \mu_M - \mu_W > 10$; $z = 1.80 > 1.64$, result significant at 5% level, accept H_1.

3 (i) 0.525   (ii) (0.501, 0.549)   (iii) 9580

4 (i) 0.978   (ii) 0.022; $62500T^2 - 1000960.4T + 4000000 = 0$; (7.66, 8.36)

5 (i)(a) 0.345   (b) 11   (c) (1.80, 2.60)   (ii) (7.42, 7.58)

Exercise 10.8

1 (i) $P(X \geq 15) = 0.083 > 0.05$, accept H_0 at 5% level
(ii) $P(X \geq 118) = 0.04 < 0.05$, reject H_0 at 5% level

2 $P(X \leq 3) = 0.011 < 0.02$, reject H_0 at 2% level; 61

3 a   (24.11, 25.09), consistent
   b   (i) $H_0: \mu = 25$, $H_1: \mu < 25$ (ii) 0.1151   (iii) 0.7257

4 (i) 0.0985, 0.00666   (ii) $z = -0.22 > -1.96$, accept H_0 at 5% level
(iii) 12, 27, 29, 19, 9, 3, 1

5 $H_0: \mu_A = \mu_B$, $H_1: \mu_A > \mu_B$; $z = 2.87 > 2.33$, reject H_0 at 1% level. The course has been beneficial in improving the mean reading score. Since the samples are not random ones, the previous conclusion is no longer valid.

Exercise 10.9

1 (i) (2.96, 3.84)   (ii) (-0.66, -0.64)

2 (i) $y = 3.24 + 4.965x$ (ii) (4.88, 5.05)

3 $y = 31.1 + 2.8\,(x - 1963)$; inappropriate as the rate of increase of wine consumption is increasing

4 $y = 51.2 + 0.032x$ (i) (50.92, 51.48)   (ii) (0.226, 0.414)

5 (i) 40.1, -0.3   (ii)(a) (-0.50, -0.10)   (b) (25.53, 30.69)
(iii) preferable because its width is smaller

Miscellaneous Exercise 10

1 (ii) £7320, £6750 (or £6792)   (iii) median not distorted by atypical high salaries or mean takes into account all the salaries

2 (i) 6.2, 19.16   (ii) £920, £438

3 (i) 1, 60; independence relevant in the calculation of the variance   (ii) ± 2

4 (i) A and B not independent   (ii) 0.25

5 (i) 0.5   (ii) 0.625   **6** (i) $\dfrac{2}{3}$ (ii) $\dfrac{19}{24}$

7 (i) 0.010495 (ii) 0.0476 (iii) under 5% of those giving a positive response to the test actually have the disease

8 a (i) $\dfrac{1}{6}$ (ii) $\dfrac{1}{12}$ (iii) $\dfrac{2}{3}$ **b** $\dfrac{7}{12}$

9 (i) 0.2787 (ii) 0.0199

10

y_i	2	4	6
p_i	$\dfrac{1}{6}$	$\dfrac{1}{3}$	$\dfrac{1}{2}$

$E[Y] = \dfrac{14}{3}$

11 (i) $\dfrac{1}{13}$ (ii) 2, $\dfrac{12}{13}$

12 (ii) 0.52, 0.64

13 (i) $\dfrac{3}{4}, \dfrac{3}{80}$ (ii) $\dfrac{3}{2}, \dfrac{3}{4}$

14 3, 3.4

15 $\dfrac{128}{9}, \dfrac{16}{9}$

16 (ii) 0.46

17 (ii) 2.4 (iii) $G(y) = \dfrac{y}{8}, 0 \leqslant y \leqslant 8$

(iv)

n_i	2	3	4
p_i	$\dfrac{5}{32}$	$\dfrac{16}{32}$	$\dfrac{11}{32}$

$\dfrac{51}{16}$

18 (i) $\dfrac{5}{3}, \dfrac{2}{3}$ (ii) $\dfrac{2}{3}$ (iii) $G(y) = \dfrac{(25y - 1)}{24}, \dfrac{1}{25} \leqslant y \leqslant 1, g(y) = \dfrac{25}{24}$.

19 (i) 0.087 (ii) 0.157 (iii) £38·40

20 (i) 0.558 (ii) 13, 11

21 (i) 7 (ii)

y_i	0	1	2	3	4	5
p_i	0.030	0.106	0.185	0.216	0.188	0.275

3.25

22 2.06; mean $\approx$ variance; 14, 27, 27, 18, 9, 4, 1, 0; $H_0: \mu = 2, H_1: \mu > 2$; 0.034

23 (i) 0.00992 (ii) 1151 (iii) 0.5940 (iv) 0.046

24 a (i) 0.099 (ii) 0.984 (iii) 0.616 (Poisson)
   b $H_0: p = 0.95, H_1: p < 0.95, z = -0.92 > -1.64$, result not significant at 5% level, retain H_0

25 (i)(a) 0.27648 (b) 0.0864 (ii) 0.8327 (iii)(c) 0.4962 (d) 0.1561 (iv) 0.081

26 (ii)(a) $\dfrac{11}{24}$ (b) $\dfrac{8}{11}$

27

x_i	0	1	2	3
p_i	$\dfrac{10}{56}$	$\dfrac{30}{56}$	$\dfrac{15}{56}$	$\dfrac{1}{56}$

(ii) $\dfrac{3}{8}$ (iii) $\dfrac{8}{13}$

28 (i) 0.480 (ii) 0.403

29 (i) 0.106 (ii) 219

30 (i)(a) 0.919 (b) 0.997 (ii) 41083 (iii) 0.822

31 (ii)

		x	
	0	1	2
y 0	0.09	0.3	0.1
1	0.18	0.3	0
2	0.03	0	0

(iii) 0.54 (iv) 6.68, 1.6176

STATISTICS 2

32 (i) 1.4   (iii) 0.16   (iv) not independent

33 1.051, 0.0256;   (1.037, 1.065); $H_0: \mu_A = \mu_B, H_1: \mu_A \neq \mu_B$; $z = 2.10 > 1.96$, reject H_0 at 5% level, conclude $\mu_A > \mu_B$.

34 (i) $y = 14.964 + 0.012x$ (ii) (0.011, 0.013)   (iii) (0.055, 0.065)

35 $y = 57.2 - 0.48x$ (i) (47.15, 48.05)   (ii) ($-10.64, -8.56$)

36 (i) 0.06, 0.132   (ii) (1.344, 1.416)   (iii) (22.10, 26.04)

37 $y = 23 - 0.325x$ (i) (19.64, 26.36)   (ii) ($-0.407, -0.243$)   (iii) (5.69, 7.81)

38 $\dfrac{\sigma^2}{[5(2 + 3k)]}$, $\dfrac{(2k + 3)\sigma^2}{125k}$; $\dfrac{2}{(3k + 2)}$

39 $\dfrac{\mu e^{\mu}}{(e^{\mu} - 1)}$

40 0.72, 1.64

41 29, 6.41   (ii) 162

42 45 km/h, 7.81 km/h

43 (i) 0.6, 0.155   (ii) 80.4, 98.0;   (74.2, 86.6)

44 **a** (i) 10.8%   (ii) 0.404   **b** 6

45 $\dfrac{\theta}{2}, \dfrac{\theta^2}{12}$; $N\left(\theta, \dfrac{\theta^2}{3n}\right)$; (i) 21, $\dfrac{2}{7}$ (ii) (18.6, 23.4)

46 (i) 0.49   (iii) £18, 45 £^{2} (iv) 0.03   (v) 0.25

47 $H_0: p = 0.5$, $H_1: p > 0.5$, $\alpha = 0.019$;
$H_0: p = 0.75$, $H_1: p > 0.75$, $z = 2.63 > 2.33$,
result significant at 1% level. Accept coach's claim.

48 (i) 0.067   (ii) 0.057
$H_0: \mu_D = 100$, $H_1: \mu_D \neq 100$, $t = 1.68 < 2.13$,
result not significant at 5% level. Retain H_0.

49 (i) 7.865 min   (ii) 7.5 min, 34.6p

50 **a** $\frac{1}{4}$ **c** 0.75, (0.662, 0.838)   **d** (i) 290   (ii) 385

51 (i) 30.2, 29.75, 4, 8.5
(ii) $H_0: \mu_1 = \mu_2$, $H_1: \mu_1 > \mu_2$; $z = 1.86 > 1.64$, reject H_0 in favour of H_1, sufficient evidence at the 5% significance level to support B's claim.
(iii) ($-0.024, 0.924$); this does not discredit A's assertion because the interval contains the value 0.

52 (i) $y = 19.25 + 1.53x$
(ii) $z = 2.16 > 1.96$, reject H_0 in favour of H_1, conclude $\beta > 1.52$

Statistical tables

Table 1

Cumulative binomial probabilities

The tabulated value is $P(X \leqslant r)$, where X has a binomial distribution with parameters n and p.

	$p =$	0.05	0.10	0.15	0.20	0.25	0.30	0.35	0.40	0.45	0.50
$n = 5$	$r = 0$	0.774	0.590	0.444	0.328	0.237	0.168	0.116	0.078	0.050	0.031
	1	0.977	0.919	0.835	0.737	0.633	0.528	0.428	0.337	0.256	0.187
	2	0.999	0.991	0.973	0.942	0.896	0.837	0.765	0.683	0.593	0.500
	3	1.000	1.000	0.998	0.993	0.984	0.969	0.946	0.913	0.869	0.813
	4			1.000	1.000	0.999	0.998	0.995	0.990	0.982	0.969
$n = 10$	$r = 0$	0.599	0.349	0.197	0.107	0.056	0.028	0.013	0.006	0.003	0.001
	1	0.914	0.736	0.544	0.376	0.244	0.149	0.086	0.046	0.023	0.011
	2	0.988	0.930	0.820	0.678	0.526	0.383	0.262	0.167	0.100	0.055
	3	0.999	0.987	0.950	0.879	0.776	0.650	0.514	0.382	0.266	0.172
	4	1.000	0.998	0.990	0.967	0.922	0.850	0.751	0.633	0.504	0.377
	5		1.000	0.999	0.994	0.980	0.953	0.905	0.834	0.738	0.623
	6			1.000	0.999	0.996	0.989	0.974	0.945	0.898	0.828
	7				1.000	1.000	0.998	0.995	0.988	0.973	0.945
	8						1.000	0.999	0.998	0.995	0.989
	9							1.000	1.000	1.000	0.999
$n = 20$	$r = 0$	0.358	0.122	0.039	0.012	0.003	0.001	0.000	0.000		
	1	0.736	0.392	0.176	0.069	0.024	0.008	0.002	0.001	0.000	
	2	0.925	0.677	0.405	0.206	0.091	0.035	0.012	0.004	0.001	0.000
	3	0.984	0.867	0.648	0.411	0.225	0.107	0.044	0.016	0.005	0.001
	4	0.997	0.957	0.830	0.630	0.415	0.238	0.118	0.051	0.019	0.006
	5	1.000	0.989	0.933	0.804	0.617	0.416	0.245	0.126	0.055	0.021
	6		0.998	0.978	0.913	0.786	0.608	0.417	0.250	0.130	0.058
	7		1.000	0.994	0.968	0.898	0.772	0.601	0.416	0.252	0.132
	8			0.999	0.990	0.959	0.887	0.762	0.596	0.414	0.252
	9			1.000	0.997	0.986	0.952	0.878	0.755	0.591	0.412
	10				0.999	0.996	0.983	0.947	0.872	0.751	0.588
	11				1.000	0.999	0.995	0.980	0.943	0.869	0.748
	12					1.000	0.999	0.994	0.979	0.942	0.868
	13						1.000	0.998	0.994	0.979	0.942
	14							1.000	0.998	0.994	0.979
	15								1.000	0.998	0.994
	16									1.000	0.999
	17										1.000

Table 2

Cumulative Poisson probabilities

The tabulated value is $P(X \leqslant r)$, where X has a Poisson distribution with mean a.

$a =$	0.5	1.0	1.5	2.0	2.5	3.0	3.5	4.0	4.5	5.0
$r = 0$	0.607	0.368	0.223	0.135	0.082	0.050	0.030	0.018	0.011	0.007
1	0.910	0.736	0.558	0.406	0.287	0.199	0.136	0.092	0.061	0.040
2	0.986	0.920	0.809	0.677	0.544	0.423	0.321	0.238	0.174	0.125
3	0.998	0.981	0.934	0.857	0.758	0.647	0.537	0.433	0.342	0.265
4	1.000	0.996	0.981	0.947	0.891	0.815	0.725	0.629	0.532	0.440
5		0.999	0.996	0.983	0.958	0.916	0.858	0.785	0.703	0.616
6		1.000	0.999	0.995	0.986	0.966	0.935	0.889	0.831	0.762
7			1.000	0.999	0.996	0.988	0.973	0.949	0.913	0.867
8				1.000	0.999	0.996	0.990	0.979	0.960	0.932
9					1.000	0.999	0.997	0.992	0.983	0.968
10						1.000	0.999	0.997	0.993	0.986
11							1.000	0.999	0.998	0.995
12								1.000	0.999	0.998
13									1.000	0.999
14										1.000

$a =$	5.5	6.0	6.5	7.0	7.5	8.0	8.5	9.0	9.5	10.00
$r = 0$	0.004	0.002	0.002	0.001	0.001	0.000	0.000	0.000	0.000	0.000
1	0.027	0.017	0.011	0.007	0.005	0.003	0.002	0.001	0.001	0.000
2	0.088	0.062	0.043	0.030	0.020	0.014	0.009	0.006	0.004	0.003
3	0.202	0.151	0.112	0.082	0.059	0.042	0.030	0.021	0.015	0.010
4	0.358	0.285	0.224	0.173	0.132	0.100	0.074	0.055	0.040	0.029
5	0.529	0.446	0.369	0.301	0.241	0.191	0.150	0.116	0.089	0.067
6	0.686	0.606	0.527	0.450	0.378	0.313	0.256	0.207	0.165	0.130
7	0.809	0.744	0.673	0.599	0.525	0.453	0.386	0.324	0.269	0.220
8	0.894	0.847	0.792	0.729	0.662	0.593	0.523	0.456	0.392	0.333
9	0.946	0.916	0.877	0.830	0.776	0.717	0.653	0.587	0.522	0.458
10	0.975	0.957	0.933	0.901	0.862	0.816	0.763	0.706	0.645	0.583
11	0.989	0.980	0.966	0.947	0.921	0.888	0.849	0.803	0.752	0.697
12	0.996	0.991	0.984	0.973	0.957	0.936	0.909	0.876	0.836	0.792
13	0.998	0.996	0.993	0.987	0.978	0.966	0.949	0.926	0.898	0.864
14	0.999	0.999	0.997	0.994	0.990	0.983	0.973	0.959	0.940	0.917
15	1.000	0.999	0.999	0.998	0.995	0.992	0.986	0.978	0.967	0.951
16		1.000	1.000	0.999	0.998	0.996	0.993	0.989	0.982	0.973
17				1.000	0.999	0.998	0.997	0.995	0.991	0.986
18					1.000	0.999	0.999	0.998	0.996	0.993
19						1.000	1.000	0.999	0.998	0.997
20								1.000	0.999	0.998
21									1.000	0.999
22										1.000

Table 3

The normal distribution

a *Distribution Function.* The tabulated value is
$\Phi(z) = P(Z \leqslant z)$, where Z is the standardised
normal random variable, $N(0, 1)$.

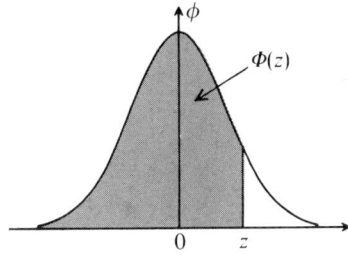

z	.00	.01	.02	.03	.04	.05	.06	.07	.08	.09
.0	.5000	.5040	.5080	.5120	.5160	.5199	.5239	.5279	.5319	.5359
.1	.5398	.5438	.5478	.5517	.5557	.5596	.5636	.5675	.5714	.5753
.2	.5793	.5832	.5871	.5910	.5948	.5987	.6026	.6064	.6103	.6141
.3	.6179	.6217	.6255	.6293	.6331	.6368	.6406	.6443	.6480	.6517
.4	.6554	.6591	.6628	.6664	.6700	.6736	.6772	.6808	.6844	.6879
.5	.6915	.6950	.6985	.7019	.7054	.7088	.7123	.7157	.7190	.7224
.6	.7257	.7291	.7324	.7357	.7389	.7422	.7454	.7486	.7517	.7549
.7	.7580	.7611	.7642	.7673	.7704	.7734	.7764	.7794	.7823	.7852
.8	.7881	.7910	.7939	.7967	.7995	.8023	.8051	.8078	.8106	.8133
.9	.8159	.8186	.8212	.8238	.8264	.8289	.8315	.8340	.8365	.8389
1.0	.8413	.8438	.8461	.8485	.8508	.8531	.8554	.8577	.8599	.8621
1.1	.8643	.8665	.8686	.8708	.8729	.8749	.8770	.8790	.8810	.8830
1.2	.8849	.8869	.8888	.8907	.8925	.8944	.8962	.8980	.8997	.9015
1.3	.9032	.9049	.9066	.9082	.9099	.9115	.9131	.9147	.9162	.9177
1.4	.9192	.9207	.9222	.9236	.9251	.9265	.9279	.9292	.9306	.9319
1.5	.9332	.9345	.9357	.9370	.9382	.9394	.9406	.9418	.9429	.9441
1.6	.9452	.9463	.9474	.9484	.9495	.9505	.9515	.9525	.9535	.9545
1.7	.9554	.9564	.9573	.9582	.9591	.9599	.9608	.9616	.9625	.9633
1.8	.9641	.9649	.9656	.9664	.9671	.9678	.9686	.9693	.9699	.9706
1.9	.9713	.9719	.9726	.9732	.9738	.9744	.9750	.9756	.9761	.9767
2.0	.9772	.9778	.9783	.9788	.9793	.9798	.9803	.9808	.9812	.9817
2.1	.9821	.9826	.9830	.9834	.9838	.9842	.9846	.9850	.9854	.9857
2.2	.9861	.9864	.9868	.9871	.9875	.9878	.9881	.9884	.9887	.9890
2.3	.9893	.9896	.9898	.9901	.9904	.9906	.9909	.9911	.9913	.9916
2.4	.9918	.9920	.9922	.9925	.9927	.9929	.9931	.9932	.9934	.9936
2.5	.9938	.9940	.9941	.9943	.9945	.9946	.9948	.9949	.9951	.9952
2.6	.9953	.9955	.9956	.9957	.9959	.9960	.9961	.9962	.9963	.9964
2.7	.9965	.9966	.9967	.9968	.9969	.9970	.9971	.9972	.9973	.9974
2.8	.9974	.9975	.9976	.9977	.9977	.9978	.9979	.9979	.9980	.9981
2.9	.9981	.9982	.9982	.9983	.9984	.9984	.9985	.9985	.9986	.9986
3.0	.9987	.9987	.9987	.9988	.9988	.9989	.9989	.9989	.9990	.9990
3.1	.9990	.9991	.9991	.9991	.9992	.9992	.9992	.9992	.9993	.9993
3.2	.9993	.9993	.9994	.9994	.9994	.9994	.9994	.9995	.9995	.9995
3.3	.9995	.9995	.9995	.9996	.9996	.9996	.9996	.9996	.9996	.9997
3.4	.9997	.9997	.9997	.9997	.9997	.9997	.9997	.9997	.9997	.9998

b *Upper Percentage Points.* The tabulated value
is z_p, where $P(Z > z_p) = p$, so that
$1 - \Phi(z_p) = p$.

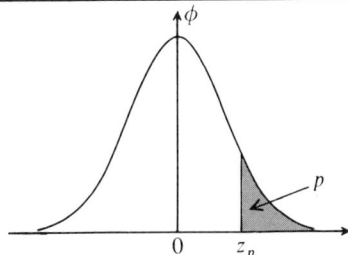

p	0.05	0.025	0.01	0.005	0.001	0.0005
z_p	1.64	1.96	2.33	2.58	3.09	3.29

231

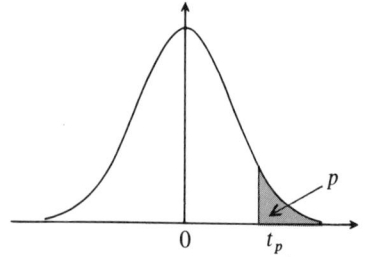

Table 4

Upper percentage points for the *t*-distribution

The tabulated value is t_p, where $P(X > t_p) = p$, when X has the
t-distribution with v degrees of freedom.

p	0.05	0.025	0.01	0.005	0.001	0.0005
$v = 1$	6.31	12.71	31.82	63.66	318.3	636.6
2	2.92	4.30	6.96	9.92	22.33	31.60
3	2.35	3.18	4.54	5.84	10.21	12.92
4	2.13	2.78	3.75	4.60	7.17	8.61
5	2.02	2.57	3.36	4.03	5.89	6.87
6	1.94	2.45	3.14	3.71	5.21	5.96
7	1.89	2.36	3.00	3.50	4.79	5.41
8	1.86	2.31	2.90	3.36	4.50	5.04
9	1.83	2.26	2.82	3.25	4.30	4.78
10	1.81	2.23	2.76	3.17	4.14	4.59
12	1.78	2.18	2.68	3.05	3.93	4.32
14	1.76	2.14	2.62	2.98	3.79	4.14
16	1.75	2.12	2.58	2.92	3.69	4.01
18	1.73	2.10	2.55	2.88	3.61	3.92
20	1.72	2.09	2.53	2.85	3.55	3.85
25	1.71	2.06	2.48	2.79	3.45	3.72
30	1.70	2.04	2.46	2.75	3.39	3.65
40	1.68	2.02	2.42	2.70	3.31	3.55
60	1.67	2.00	2.39	2.66	3.23	3.46
120	1.66	1.98	2.36	2.62	3.16	3.37
∞	1.64	1.96	2.33	2.58	3.09	3.29

Table 5

Random digits

The tabulated digits are independent observations from a distribution in which the digits 0–9 have equal probabilities.

90159	96465	96858	73429	74257	19970	87916	89204	08631	82959
63835	94333	80384	57737	41936	82487	55651	58671	72611	03774
83793	02044	76074	78034	19256	29922	56192	86943	19716	27732
37954	93470	67333	52645	49638	02802	74126	57816	68432	06050
52203	80860	02071	28397	13988	16152	63299	62855	93960	03357
69684	84874	19256	74311	14029	66457	64647	98847	01753	43677
41089	71096	43777	31817	58684	36247	04775	88407	80496	32094
03015	41571	08057	85319	09346	17075	03058	91232	23473	95982
77980	81547	74092	32879	60549	30487	02476	43878	80824	90470
50512	38559	17847	31353	01305	67204	47248	65847	39103	78374
49195	72090	22959	61455	14242	08028	09064	88436	19695	92076
81070	06070	02688	30084	03248	56913	46961	11143	67229	96523
00237	39550	75537	04273	43291	69091	03682	14784	50468	67799
43764	21488	62091	20761	69330	07661	61564	75202	01854	36385
10350	16762	67504	11431	87820	99979	84539	98135	83516	02353
25163	41056	64314	87456	55548	62706	03370	01131	84842	68637
17533	12864	00959	95443	54257	97194	30811	98350	44864	76917
16187	08144	03848	65933	88249	90334	31860	09413	11770	63679
66049	15483	80910	09046	83696	73702	32145	30374	42841	47797
25002	07562	11250	27858	72007	85400	55906	12765	31490	91355

Table 6

Random observations from the standardised normal distribution N(0, 1)

0.593	−0.878	0.837	−0.273	−0.279	−1.007	−2.119	0.048	0.941	0.257
0.035	−1.604	−0.303	−1.315	1.459	−0.555	0.765	0.544	−1.134	−0.067
0.242	−0.882	−1.549	1.627	0.436	−2.704	−0.528	−0.271	−0.732	0.005
0.020	1.185	0.970	−2.011	−0.469	−0.035	0.754	−0.265	−0.350	−0.532
0.289	−0.589	−0.475	−1.015	1.420	0.009	0.845	0.702	0.191	1.556
−1.860	0.716	−1.222	0.166	0.816	0.258	0.406	0.189	1.339	−1.840
−0.947	0.794	0.868	0.189	0.094	−0.148	−1.725	−0.224	0.381	−0.466
0.483	0.374	−0.531	1.641	0.457	0.067	0.291	0.534	0.086	−0.195
1.171	−0.387	1.679	−0.655	2.127	0.199	0.610	2.017	−0.692	0.008
0.432	−0.741	−0.583	0.730	0.598	0.985	−0.368	0.208	−0.965	0.027
1.087	−0.522	−1.153	0.038	−0.943	0.467	−0.422	−0.758	−2.175	0.480
−0.398	−0.025	−1.635	0.265	0.314	−1.398	−0.981	0.687	−0.949	1.228
−0.942	0.020	−0.063	−2.387	0.111	1.463	0.183	−0.038	−1.170	−0.202
1.652	−0.729	−0.329	1.709	−1.532	−0.855	0.290	2.079	0.005	−0.219
−0.815	0.030	−0.577	−0.038	−0.022	0.860	1.569	−1.190	−2.254	−2.697
−0.531	1.080	0.461	0.053	−0.763	0.297	1.200	0.605	−0.228	−2.191
−0.868	0.919	0.807	0.543	−2.395	−0.679	0.223	0.267	−0.414	−0.044
−1.226	−0.388	−1.174	1.566	0.332	0.560	−0.372	−1.010	0.312	1.673
−0.918	−0.709	0.096	0.407	−0.145	0.134	−0.655	0.483	−0.744	−0.359
0.284	0.761	1.836	−0.468	−1.357	−0.630	−0.101	−1.057	−0.972	1.080

Index